Biodiversity

Social & Ecological Perspectives

Vandana Shiva
Patrick Anderson
Heffa Schücking
Andrew Gray
Larry Lohmann
David Cooper

Zed Books Ltd.
London and New Jersey

World Rainforest Movement
Penang, Malaysia

Biodiversity was published by
Zed Books Ltd.
7 Cynthia Street, London N1 9JF, UK and
165 First Avenue, Atlantic Highlands, New Jersey 07716, USA,
and by World Rainforest Movement
228 Macalister Road, 10400 Penang, Malaysia,
in 1991.

Second impression, 1995.

ISBN 1-85649-053-X Hb
ISBN 1-85649-054-8 Pb

A catalogue record for this book is available from the British Library.
US CIP is available from the US Library of Congress.

Printed by Jutaprint
54 Kajang Road, 10150 Penang, Malaysia.

ABOUT THE AUTHORS

Vandana Shiva, a physicist and philosopher, is Director of the Research Foundation for Science, Technology and Natural Resource Policy, Dehra Dunn, India. She is the author of *Staying Alive, The Violence of the Green Revolution,* as well as *Ecological Audit of Eucalyptus Cultivation.* She is also the Science and Environment Advisor of the Third World Network.

Patrick Anderson became involved in rainforest conservation in 1979, when he joined a logging blockade in one of the last subtropical rainforests in Australia. The area now forms part of the Nightcap National Park and is a World Heritage Site. From 1984 to 1990, Patrick worked with the Rainforest Information Centre, coordinating its campaign to reform Australia's development assistance programme. He now coordinates Greenpeace's tropical forests campaign.

Heffa Schücking, born in 1959, was alerted to the crisis of tropical deforestation through her studies in primatology at Bielefeld University. In 1987, she helped found the Working Group for Rainforests and Species Conservation (ARA) to bring both a biological and political perspective to the rainforest movement in Germany. Heffa is the German representative for the Gaia Foundation, and works to have the interests and perspectives of forest peoples recognised by the German government and public.

Andrew Gray was trained as a Social Anthropologist at Edinburgh and Oxford Universities. He has spent two years living with the Amarakaeri people of the Peruvian Amazon. Between 1983 and 1989, he was an Executive Director of the International Work Group for Indigenous Affairs in Copenhagen. He is currently working with indigenous peoples in a collaborative research programme of IWGIA, looking into self-determination, development and identity among indigenous peoples.

Larry Lohmann taught at universities in the US and Thailand in the 1970s and 1980s, then worked for several years in Bangkok with Thai Volunteer Service and the Project for Ecological Recovery. He is now based at *The Ecologist* magazine in the UK.

David Cooper is an agricultural scientist working with Genetic Resources Action International (GRAIN), an international independent NGO promoting world-wide campaign for popular control over plant genetic resources. He is a member of Council, and former Secretary of the London-based World Development Movement.

CONTENTS

INTRODUCTION

The World Rainforest Movement (WRM), an international coalition of grassroots movements, was formed in 1987 as a response to the growing crisis of survival of forest communities.

For us, it was clear that the crisis of survival of tribal and peasant societies was a crisis of diversity. The disappearance of ecological diversity and the intolerance of cultural diversity are processes that were intimately linked. The survival struggles of different communities are in fact struggles for the preservation of diversity.

At the end of the 1980s, "biodiversity" suddenly appeared as a buzzword in conservation circles. It was given concrete expression in the World Research Institute (WRI), World Bank, IUCN, and World Wildlife Fund (WWF) publication, *Conserving the World's Biological Diversity*.

We at the WRM are troubled at the way the issue of biodiversity conservation was handled in this document and earlier drafts.

While the crisis of biodiversity erosion is focused as an exclusively tropical and Third World phenomena, the thinking and planning of biodiversity conservation is projected as a monopoly of institutes and agencies based in and controlled by the industrial North. It is as if the mind is in the North, the matter is in the South; the solution is in the North, the problems in the South. This polarity and dualism underlie the basic shortcomings of the monograph under review. Instead of being titled *Conserving the World's Biological Diversity*, the report could more honestly have been titled "The North Conserving the South's

Biological Diversity For Itself".

It is, of course, true that the tropics is the cradle of the planet's biological diversity, and that the multiplicity and variability of ecosystems and species that exist there is incomparable. However, the crisis of the erosion of diversity is a crisis in the North as much as in the South. I would, in fact, go further and state that the roots of the crisis of diversity in the South lie in the North. Both these aspects of the destruction of diversity are, however, not addressed in the analysis by WRI *et al.*

Closely related to the total neglect of forces and factors in the North as part of the problem is the neglect of the crisis of diversity in what are viewed as "production" spheres – of forestry, livestock, and agriculture. Among the proximate causes identified as leading to the loss of biological resources are clearing and burning of forests, overharvesting of plants and animals, and indiscriminate use of pesticides. However, in the last twenty to thirty years, in addition to these factors, the deliberate substitution of diversity by uniformity – of crops, trees, and livestocks, through development projects financed by aid from international agencies – has worsened the biodiversity crisis.

The World Bank which has now launched the Biodiversity Action Plans, has for the past forty years been financing the destruction of genetic diversity in the Third World. It has financed the Green Revolution which replaced genetically diverse indigenous cropping systems in the Third World with vulnerable genetically uniform monocultures.

It has contributed to genetic erosion through the centralised research institutions controlled by the Consultative Group on International Agricultural Research (CGIAR). The CGIAR was launched in 1970 at the initiative of the World Bank and has its offices in the World Bank.

In fact, the continued spread of genetic uniformity is perversely viewed as a means for "biodiversity conservation" in programmes of the World Bank. As an illustration, John Spears, of the World Bank has recommended the intensification of monoculture practices in agriculture and forestry to 'preserve the biological diversity' in Asia. This schizophrenic approach to biodiversity which adopts a policy of destruction of diversity in production processes, and a policy of preservation in 'set-asides' cannot be effective in the conservation of species diversity. Biodiversity cannot be conserved unless production itself is based on a policy of preserving diversity.

Thus, World Bank-financed 'social-forestry' projects in India are, in effect, the replacement of genetically diverse agro-forestry species and crops with monocultures of eucalyptus plantations for the paper and pulp industry. The globally launched Tropical Forestry Action Plan (TFAP) is similarly leading to the spread of genetic erosion worldwide.

Probably the treatment of Northern agencies only as part of the solution, not as part of the problem, is related to the economistic paradigm held by the authors of the report. In the chapter on the "Values of Biological Diversity", it is recognised that biological resources have social, ethical, cultural and economic values. "But", the authors proceed to say; "in order to compete for the attention of government decision makers in today's world, policies regarding biological diversity first need to demonstrate in economic terms the value of biological resources in economic terms to a country's social and economic development".

The economic values of biological resources are then divided into:

1. **Consumptive use value** – value of products consumed directly without passing through the market such as firewood, fodder, and game meat;

2. **Productive use value** – value of products commercially exploited; and

3. **Non-consumptive use value** – indirect values of ecosystem func-

tions, such as watershed protection, photosynthesis, regulation of climate and soil-production.

An interesting value framework has thus been constructed which predetermines analysis and options. If the Third World poor, who derived their livelihood directly from nature only "consume", and the trading and commercial interests are the only "producers", it follows quite naturally that the Third World is responsible for the destruction of its biological wealth, and the North alone has the capacity to conserve it.

This ideologically constructed divide between consumption, production and conservation hides the political economy of the processes which underlie the destruction of biological diversity.

Third World peasants and forest dwellers **participate** in the ecological processes of reproduction of biodiversity. They do not "consume" it in the sense of a one time destructive use. And it is in participating in the biological renewal of life, through their cultures and lifestyles, that they simultaneously achieve 'production' and 'conservation', categories which in industrial civilisation are separate and opposed. These separated and dualistic categories underlie the approach of *Conserving the World's Biological Diversity*, and this is its major shortcoming when viewed from the standpoint of the South, especially the marginalised cultures of the South whose lives are most intimately linked to diversity.

Contrary to the false categorisations of dominant conservation programmes, it is production systems based on uniformity and commodification which "consume" biological resources in the original sense which meant to destroy, to pillage, to subdue, to exhaust. Unless these categories of production and consumption are switched around, the conservers of the world's biological diversity will be treated as those who threaten it, and the predatory systems that "consume" biodiversity as raw material to the point of extinction will be perceived as essential to programmes of conservation. We would like to turn this logic upside down and inside out through the contributions in this book which attempt to articulate issues of conservation from the viewpoint of those who practise and preserve diversity.

Heffa Schücking and Patrick Anderson in *Voices Unheard and Unheeded* analyse the dominant paradigm of conservation and reflect on its lacunae. Andrew Gray's contribution is based on his experience with indigenous peoples of Latin America. Larry Lohmann's article, *Who Defends Biological Diversity?*, reflects the conservation strategies of the peasant and forest dwelling communities in the case of Thailand. David Cooper provides a critique of the global biodiversity convention and suggests the vital components needed for a meaningful agreement. My own piece attempts to show that production based on principles of uniformity is the biggest threat to biodiversity. This threat will increase with the emergence of new biotechnologies which will erode biological diversity, both through acceleration of the introduction of uniformity in production as well as through ownership of life forms as private property.

Reflecting the philosophy of diversity and pluralism of the WRM, each contribution takes its own approach and is reproduced unedited in this collection. We hope these essays will inspire debate and reflection and will contribute to a richer understanding of the issues involved in the conservation of biodiversity.

We would like to thank the WRM secretariat, especially Nora Syed Ibrahim for editorial assistance and Christina Lim for the book design, Lim Jee Yuan for the cover design , and Linda Ooi for the production of this book.

Vandana Shiva.

Voices Unheard and Unheeded

Heffa Schücking and Patrick Anderson

"I can still hear them calling, but our groups have been cut off from each other for some time now. Ever since the road was cut through our home, and more and more humans have come into the forest. We are only six on our side now and have retreated to the last quiet places.

Yesterday I went to the edge. Everything is destroyed there, and from all sides the wasteland is growing. Our forest is dying and there is no place we can go. When we woke this morning we couldn't hear calls from the other side anymore. Across the road the forest is burning...."

The last representatives of the black-faced lion tamarin cannot tell us whether or in what way they perceive their threatened extinction. The species was discovered by scientists in June 1990 on a small island off the east coast of Brazil. The sensational discovery of this new monkey species with the scientific name, *Leontopithecus caissara*, in a densely populated area, is overshadowed by the fact that only one dozen individuals could be found. The last members of this cat-sized monkey species live in coastal forest areas of the island which are being rapidly developed for tourism (Conservation International, 1990). *Leontopithecus caissara* is one of the 26 primate species that will probably become extinct within the next ten years.

A mass extinction of species is not the sort of event that one would expect to pass unnoticed by most people. But just that is happening. We are not hearing the voices of the dying species, as we pay no attention to the myriad life forms with which we share the planet. Especially those of us living in urban societies, estranged from the natural world, have lost sight of the fact how narrowly defined the environmental conditions are that enable the survival of all of the Earth's species, including our own.

The Biodiversity Crisis

To date, 1.4 million life-forms have been named and described by science. Biologists estimate that there are at least five million, but perhaps as many as fifty million species existing today (McNeely *et al.*, 1990). The estimate has increased dramatically in recent years, following research in the rainforest canopy and especially due to the work of tropical entomologist, Terry Erwin. If his research areas are representative, the tropical rainforests probably support some 30 million insect species. Other species-rich groups such as fungi and mites have also barely been investigated, so that we are still unable to estimate the total number of species to the nearest order of magnitude (Wilson, 1990).

Plant and animal species from all the major families are still being discovered in the rainforests. During the last five years over 300 new fish species have been classified from the Amazon region. Even among the primates, who belong to the most conspicuous and well-researched animals in the rainforests, four new species were discovered during the last few years. Botanist Alvin Gentry estimates that between 15,000 and 20,000 unknown species of tropical flowering plants exist. At present, probably less than five per cent of the biological diversity of the rainforests is known to science (McNeely *et al.*, 1990). The task of recording all of the Earth's life-forms would take the entire working lives of 25,000 taxonomists. Worldwide, there are probably fewer than 1,500 taxonomists able to undertake the work of classifying tropical rainforest organisms (Wilson, 1988). Even if we had the taxonomists we would still know next to nothing about the life-cycles of these species and their ecological relationships.

According to an estimate by Peter Raven (1989), Director of the Missouri Botanical Gardens, we may already be losing 100 species per day. E.O. Wilson (1988) and Norman Myers (1986) estimate that 50 species are being driven to extinction per day. The vast majority of these extinctions are occurring in tropical forest areas. The differences between these and other estimates follow from different assumptions of the total number of species and ecosystem destruction rates. Wilson acknowledges that his estimate is conservative, as it is based on the assumption of 5 million rainforest species in total, and a tropical deforestation rate of 0.7 per cent per annum (recent evidence indicates the rate is three times as high). Despite the variation in different estimates, all predictions lead to the conclusion that what is taking place today is more than the loss of individual species. For the first time in human history, biodiversity as a whole is endangered. Conservation biologists warn that 25 per cent of all species could become extinct during the next twenty to thirty years.

The End Of Natural Evolution?

Since life first evolved on Earth some four billion years ago, the extinction of species has been a fact of life. Of all species that ever lived on Earth, only between one and six percent exist today. Paleontologists estimate, for example, that the average life span of vertebrate species is about five million years. Mass extinctions took place before *Homo sapiens'* emergence on the planet; the last one being 65 million years ago, wiped out nearly all the marine plankton, many major groups of marine invertebrates and all dinosaurs. Are humans really responsible for the present mass extinction?

During the last 200 million years, something like 90 species became extinct in each century due to natural evolutionary processes (Raup, 1986). At the same time evolution brought forth new life forms that more than replaced the species that were lost. Today, the extinction rate is approximately 40,000 times higher than this "background" rate due to the activities of *Homo sapiens.* While this mass extinction threatens to surpass mass extinctions of the past, the conditions for the evolution of new species have never been worse. There are several reasons for this, which are all interconnected:

- Never before has one species influenced the environmental conditions all over the planet to such an extent. The sum of human activities, including agriculture, forest clearance, grazing and land under urban and industrial sprawl, indicate that our species presently uses forty per cent of the annual net photosynthesis production of the planet. The consumption of two fifths of annual net food resources of the planet by one species is clearly incompatible with biological diversity and stability (Vitousek *et al.*, 1986).

- Practically all ecosystems are threatened by human-caused destruction and degradation. Especially alarming however, is the loss of ecosystems that play a critical role in evolutionary processes. These areas include coral reefs, ancient lakes, ecosystems of the mediterranean climate region, tidal zones and, of course, the tropical rainforests. As so-called "power houses of evolution", these forests show the highest rate of evolutionary diversification (Myers, 1988). Together with other species-rich biomes, they are literally the "womb of life".

● For the first time, a significant proportion of terrestrial plant species which forms the basis of land ecosystems, are threatened (Knoll, 1984). Previous mass extinctions had no significant effect on terrestrial plants, but today, one fifth of all plant species on land face extinction within the next 20 years. The loss of plant species will not only drive many other organisms to extinction; it will seriously deplete the "resource base" on which evolutionary processes can operate in the future (Myers, 1986).

The impact of human influence on natural evolutionary processes is so severe that renowned biologists are referring to the current crisis not only as leading to the death of species, but also as leading to an "end of birth" (Soulé and Wilcox, 1980).

"Proximate" Causes For The Loss Of Species

The causes for the loss of species are numerous but the most important is, without question, the loss and fragmentation of natural habitats. A study in 1986 by the International Union for the Conservation of Nature and Natural Resources (IUCN) found that south of the Sahara, 65 per cent of the original ecosystems have been subject to major ecological disturbance, and for South East Asia, 67 per cent of natural habitat has been lost. In Western Europe, the percentage of habitat loss is even higher.

The over-exploitation of plant and animal species also plays an important role in the destruction of habitats. The Species Survival Commission of the IUCN has estimated that this is a decisive factor in the impending extinction of two out of five of threatened vertebrate species.

The impact of exotic species is a causal factor that is often underestimated, even by conservationists. Introduced fish species, for example, threaten to wipe out a large part of the diverse fish fauna of the big

African lakes. Roughly half of the extinctions of island bird species have been due to the introduction of exotic mammals. Goats and other introduced herbivores have devastated whole island ecosystems. Exotic plant species are now regarded as the major threat to the US National Parks system (McNeely *et al.*, 1990). Unfortunately, such facts are neglected in current discussions about the release of genetically manipulated organisms. Biologists such as Daniel Janzen (1986) fear that the release of such organisms may lead to extinction catastrophes of an unprecedented dimension.

Pesticides, tropospheric ozone, sulphur and nitrogen oxides, and other "blessings" of industrial society also play a decisive role in the degradation of natural systems and the endangerment of species. A recent study showed that 98 per cent of the conifer forests in eastern and middle Europe are threatened with rapid destruction. 84 per cent of the deciduous forests in eastern Europe, and 50 per cent in middle Europe are also severely damaged (IIASA, 1990). This 'Waldsterben' initially only leads to lower numbers of individuals within species due to the loss of suitable habitat. In the end-phase of ecosystem deterioration, a rapid decline of species inevitably follows (LÖLF, 1986).

In the future, climate change could become the main threat facing many plant and animal species. Climatic change already appears to be happening at a faster rate than ever before and this will have drastic consequences for most ecosystems. Many natural habitats are now fragmented into "islands" amidst vast seas of agricultural and industrial wastelands. Because of habitat fragmentation, even mobile species will not be able to migrate to more suitable areas in the face of climate change (McNeely *et al.*, 1990). The evolutionary consequences of changes in the composition of the Earth's atmosphere are still unknown, but present knowledge shows that many organisms are very sensitive to fluctuations in the carbon dioxide (CO_2) concentration in the atmosphere (Fajer, 1989).

Underlying Economic And Political Causes Of The Loss Of Biodiversity

To understand and be able to stop the loss of biodiversity, the "proximate" causes listed above must be linked to the driving forces generating them. In our wider analysis, we have stated that *Homo sapiens* is responsible for the current depletion of biological diversity. But this responsibility is not evenly divided. In order to address the causes of habitat loss, pollution, climate change and the over-exploitation of species, we must identify their political and economic roots as well as the forces in society that show the potential for ecological transformation.

Such a comprehensive analysis is beyond the scope of this article, so we have focussed on the issue of habitat loss, as represented by the destruction of tropical rainforests.

The socio-economic and political causes of biodiversity loss vary from region to region. In recent times, they can be linked to governmental and international support for industrial forestry, agriculture and energy programs over and above traditional usage patterns. The enormous fires in the Amazon have been fueled by two main sources; state subsidies for the cattle industry that allow totally uneconomic beef ranches to be carved out of the jungle (Caufield, 1984; Lutzenberger, 1987); and the taking over of the fertile lands in the North-East and South of Brazil by agribusiness operations to grow export crops. As Jose Lutzenberger (now the Brazilian Secretary for Environment), noted in 1989:

> "The promotion and subsidy for big mono-cultures of soya beans did not take place to feed hungry Brazilians, but to feed fat cows here in the Common Market, cows that are producing what even the technocrats of the European Market call 'seas of milk and mountains of butter'. It was this policy of promoting cash crops in the

> South that destroyed one of the most well-adapted, healthy and ecologically sane peasant cultures that we had in the South of Brazil. Hundreds of thousands of people, maybe a million, lost their lands or lost access to land, and these are the people that are now being driven into the rainforest."

In South-East Asia, it is the tropical timber industry that is cutting down the last primary rainforests, again against the will of local people (Nectoux and Kuroda, 1989; Colchester, 1989; Hong, 1986). The industry is condoned by the elite: politicians in Malaysia, and army generals in Indonesia, Burma and the Philippines. As in Brazil, the traditional practices and rights of local communities are violated, and the destruction of resources for the benefit of the few is justified in the name of national development and the generation of foreign exchange.

The picture is incomplete without including the role of western culture in this process of the usurpation of the control of local communities and the liquidation of natural resources for short term profit. As Harrison Ngau (1989), from Sahabat Alam Malaysia, noted:

> "The roots of the problem of deforestation and the waste of resources are located in the industrialized countries, where most of our resources, such as tropical timber, oil, natural gas and fish end up. The rich nations with one quarter of the world's population consume four-fifths of the world's resources. It is the throw-away culture of the industrialized countries now advertised in and forced on to Third World countries that is leading to the throwing away of the world. Such so-called `progress' has simply led to destruction and despair".

The introduction of exotic species of plants and animals causes major disturbances to ecosystems because the checks and balances that hold the equilibrium in the native ecosystem do not apply to the introduced species. In the same way, when foreign capital and technology are introduced into traditional socio-economic systems, the checks and balances that keep the activities of local communities to a sustainable level do not apply. Without these indigenous patterns and controls, such as communal decision-making and ownership, simple technology, taboos against over-exploitation, and satisfaction with a simple standard of living, exotic capital and technology can swiftly displace whole patterns of sustainable livelihood, channelling profits to a few beneficiaries. Where respect for human rights is lacking and state based environmental protection is weak, untold misery for local people and environmental despoilation are the result (Chin *et al.*, 1989).

Development assistance plays a central role in this process, due to the development models that are offered, and to the fact that most donor countries spend the majority of their assistance funding businesses from their own country (ARA/infoe, 1989). Development assistance can be likened to the AIDS virus; a pathogen that destroys the ability of the host country to resist the invasion of a foreign socio-economic system. Throughout the Third World, the result of large scale AID has been the displacement of traditional cultures and sustainable patterns of land use, along with the rapid liquidation of forests and agricultural lands for the benefit of the industrial elite (Bank Information Centre, 1990).

Commenting on the development model that has lead to the present crisis in Brazil, the Brazilian scientist Jose Goldemberg (1989), noted;

> "What has happened and is happening in the Amazon today is not only the result of action by Brazilians. The great projects that characterized the developmentalist model involved private foreign capital and the action of international agencies. The former have frequently been as destructive, if not more so, as national agencies and businesses. Furthermore, much of the Amazon's developmentalist policy is geared to export and is tied

> to impositions derived from the burden of Brazil's foreign debt to satisfy the economic interests of other countries."

If the process of biological destruction in the Third World is to be checked, these voices must be heeded. Many of the causal factors for large-scale destruction of ecosystems can be traced back to the patterns of consumption, political and economic policies as well as corporate activities of the industrialized world.

The Destruction In Ecological Comparison

The loss of biological diversity is a global crisis. There is hardly a region of the Earth that is not facing ecological catastrophes. It can therefore be asked with what justification the demise of tropical rainforests is playing such a central role in the international environmental discussion on biodiversity.

If we look at the desperate situation for nature in Europe or North America, it may seem inappropriate for the people of these regions to be so concerned about environmental problems in the tropics. Fifty per cent of the native plant and animal species in West Germany are threatened with extinction, whereas, in the rainforest currently only 10 to 15 per cent are threatened (Simberloff, 1986).

The loss of 15 per cent of the species of the tropical rainforests, however, is a completely different order of biological catastrophe than a similar level of extinctions elsewhere. A ten hectare patch of rainforest in South-East Asia may well support more tree species than the whole of North America. A biological survey of a single tree growing in Peru found it supported more species of ants than exist in the whole of Great Britain. One square kilometer of rainforest in Central or South America may be home to hundreds of bird species and tens of thousands of insect species. The tropical rainforests are home to at least 50 %, but possibly up to 90% of all species (McNeely *et al.*, 1990).

Another quality that must be taken into consideration, is the fact that many of the impending extinctions in temperate countries, are "only" a local phenomenon. If the sea eagle becomes locally extinct in Germany, it will (hopefully) still have a suitable habitat elsewhere. Extinctions in the rainforests usually mean the complete obliteration of a species. In comparison to most species in the temperate zones, most rainforest species generally have small distribution areas and low population sizes. A single agricultural or hydro-electric mega-project in a rainforest area may drive thousands of plant and animal species to extinction (ARA/infoe, 1989).

As the oldest land ecosystems, the tropical rainforests are characterized by the enormous and complex set of ecosystem interconnections. As Peter Raven has noted, "there are not only five to ten times as many species in the rainforests, as elsewhere on Earth, but these forests are characterized by between 100 and 1,000 times as many ecological interactions." Due to these intricate co-evolutionary dependencies, the sensitivity and fragility of ecological stability in tropical rainforests is of a completely different order from that in temperate forest ecosystems. In most temperate regions, if destructive forestry and agricultural practices were stopped, there are good prospects for the regeneration of natural areas. The prospects for the regeneration of rainforests are markedly poorer, especially where large areas have been destroyed (Jacobs, 1981).

The rainforests are also among the most swiftly disappearing ecosystems. In 1989, approximately 142,200 square kilometers of tropical rainforests were completely destroyed. A further 200,000 square kilometers were severely degraded. Since 1979, the rate of destruction has risen by 90 per cent (Myers, 1989). If this increase in the rate of loss is not stopped, virtually all of the remaining rainforests will be lost in the next 20 years.

It was the demise of tropical ecosystems that prompted Paul and Ann Ehrlich in the early 1980's to express their fear that while conservationists in the temperate zone were fighting a rearguard battle for a "few endangered species and populations" in their countries, "the treasury of organic diversity is being plundered in the Tropics" (Ehrlich and Ehrlich, 1981).

A Socio-Economic Assessment

In many reports and articles concerned with the crisis of biological diversity, the tropical rainforests are described as "gene banks" or "pharmacies". Regret is usually expressed that due to their ongoing destruction, the industrial world is losing the chance to utilize new "miracle" plants and important genetic resources. It is, of course, legitimate to point to the immense importance that these forests already have for agriculture, medicine and industry in the industrialized world, as well as to their future potential in this regard. But this line of argument has taken up so much room in the international discussion, that the value of the rainforests for local human populations has barely received any attention in comparison.

Human usage, not only of the rainforests, but of all ecosystems, can be analysed under three main aspects (see also McNeely *et al.*, 1990). These are:

- Ecological functions; including the maintenance and protection of watersheds and soil, the regulation of climate and habitat for wild plant and animal species.
- Subsistence values; including the provision of foods, fibres, medicines and other products that are consumed outside of a market economy.
- Commercial usage; including extractive products sold on local and

regional markets; export goods such as resins, dyes, rattan and tropical timber; and genetic resources used in agricultural and forestry crops.

Ecological Functions

If we compare these values, then the preservation of the ecological functions of the rainforests undoubtedly have the highest social and economic priority. Although tropical rainforests cover only one sixteenth of the land surface area of the Earth, they receive half of all land rainfalls. Millions of tons of water move through these forests in an endless cycle of transpiration, evaporation and precipitation. As the rainforests are destroyed, the capacity to regulate temperature and moisture is lost, severely affecting local and regional climates.

Tropical forests act as gigantic water reservoirs, protecting areas downstream from floods and droughts. Studies of the disturbance of the water regime in the Ivory Coast have shown that rivers coming from the Tai National Park have three to five times as much water at the end of the dry season as rivers from adjacent areas where the watershed forests have been destroyed for coffee plantations. Throughout the year, but especially during the dry period, this water has an essential role in downstream agriculture (Dosso *et al.*, 1981). In the Amazon, most human settlements lie along the rivers and orient their agriculture to the regular flooding cycles. The disturbance of watershed forests has led to grave losses for the farming communities downstream. In some deforested areas of Peru, water runoff is one hundred and fifty times as high as in intact forest. Downriver agricultural production in these regions has suffered heavily because of new patterns of increased flooding (Myers, 1988).

In intact rainforests, only part of the rainfall reaches the forest floor, the rest is absorbed by the vegetation. Studies in South-East Asia have found that the canopy of an intact primary rainforest can absorb up to 35 per cent of the rainfall. After logging, this rate is reduced to less than

20 per cent. In rubber plantations, the canopy typically absorbs only 12 per cent of the rainfall (Daniel and Kulasingham,1974). The decreased absorption of rainfall through the vegetation leads to an enormous increase in soil erosion. Palm oil plantations for example, may lose up to 11 times as much soil as natural forests. On agricultural land in the tropics, the erosion rate can be 34 times that for natural forest areas (Myers, 1988). The consequences include flooding, erosion, siltation, pollution of drinking water and serious decline of fish populations.

The ecological consequences of the degradation and destruction of rainforests affect many more people than the 200 million tropical forest inhabitants. 40 per cent of the farmers in developing countries live in agricultural areas that are dependent on water from rainforest watersheds (Myers, 1985). Not without reason are the rural populations in Thailand, Malaysia, Indonesia and the Philippines actively protesting deforestation. The largest group of refugees in the Third World today are the millions of people who have been displaced through the destruction of nature (Frankfurter Rundschau, 1988). Securing the ecological functions of the forest is synonymous with securing the survival of those who depend on them.

Subsistence and Local Market Values

From a social or economic perspective, the protection of the subsistence usage of rainforests is also an absolute necessity. For 62 developing countries, wild game plays an essential role in meeting the protein needs of their populations. In 19 countries wild game makes up more than 50 per cent of the meat consumed (Secrett, 1986).[1]

A study from Peninsular Malaysia showed that over 1,200 wild plant species are used by the rural population; equal to one plant in six (Jacobs, 1982). Since no medicinal and not all food plants were included in the study, the true usage rate could be as high as one plant species in three (Myers, 1988). Around the world, three billion people still rely on traditional healing practices and medicines. In the tropical countries, most of these medicines come from rainforest areas. A study of four tribes in the Amazon found that they use between 50 and 80 per cent of the tree species in their area (Prance *et al.*, 1987).

The value of tropical forests for subsistence uses is the category that has received the least recognition in the international discussion on rainforests. The subsistence usage of wild plants and animals plays almost no role in the industrialized countries today and has subsequently been ignored by so-called development strategies for the tropical forests. In the tropical forest countries however, large segments of the population are immediately dependent upon local biological resources. This is especially true of the forest peoples and the rural poor. The examples cited above show the extent of their usage of tropical forest ecosystems. When traditional community structures are still intact, these uses are generally sustainable.

Commercial Usage

It is the commercial value of rainforests that receives the most consideration internationally. The commercial value is almost always very narrowly defined and does not take account of the importance of the extractive economy. The value of extractive products sold on local markets and regional markets (such as dyes, oils, resins, honey, fibres, plant foods, wax, fire-wood, medicinal plants, fish and game) play an essential role in securing the livelihood of millions of local people (de Beer, 1990).

Neither the ecological nor the subsistence usage value of rainforests show up in the Gross National Product calculations of tropical countries. Even the commercial usage values of tropical forests are reduced to a narrow range of goods and values that have been legitimized by development planners and economists. Only a small section of the commercial usage, such as the export value of tropical timber, are usually represented. This is despite the fact that the tropical timber industry seriously degrades and destroys some 5 million hectares of tropical forests each year (World Bank, 1988), causing enormous losses to subsistence and extractive economies and seriously degrading ecological values. Tragically, such destructive concepts of commercial usage still dominate the discussion of economic values of the rainforests.

The analysis of the full range of socio-economic benefits derived from intact rainforests show that the only truly economically viable option is to preserve the remaining areas in their natural state. It also illustrates the narrow-mindedness of conservationists who attempt to justify the economic value of rainforests by refering exclusively to their value for Western agriculture, medicine and industry. A broad discussion of the priorities in our assessment of economic values is needed. All programs for the conservation of biological diversity must give priority to the preservation of ecological functions and subsistence uses over and above new schemes of commercialisation of biological diversity for the international market.

The Biodiversity Conservation Strategy Programme

The World Resources Institute (WRI), IUCN and the United Nations Environmental Programme (UNEP) are preparing a much needed global programme to address the biodiversity crisis. The Programme has three components:

- The Biodiversity Status Report; a compendium of data on the status and trends of biodiversity and actions to conserve biodiversity. It will be released at the end of 1991 and updated every two years;
- The Biodiversity Conservation Strategy, and;
- The Decade Action Plan for Conserving Biodiversity.

The Strategy and Action Programme are to be launched in early 1992, in time for consideration at the United Nations Conference on Environment and Development in Brazil. The goals of this Programme were set out in "*Gift to the Future – A Strategy for Local, National and International Action to Save, Study, and Sustainably Use the World's Biological Diversity*" (1989).

Much of the material that has been published in the preparation of this Programme, including parts of "Conserving the World's Biological Diversity" (McNeely *et al.*, 1990) is excellent. It has already received high acclaim from members of the scientific conservation community and many international conservation groups. In our opinion, many of the basic ideas of the Biodiversity Conservation Strategy Programme (BCSP) and its call for a global initiative for biodiversity conservation are useful. It also offers the chance for a long over-due exchange between members of the scientific conservation community and grass-roots organisations. Yet, there are still many fundamental questions regarding the BCSP that need to be addressed. Most of these questions are equally applicable to other activities and programmes underway to stem the loss of biodiversity.

Save It, Study It, Use It

The BCSP initially declared a three-tiered approach of "Save it, Study it, Use it" as its central motto. A long standing split exists in the conservation movement between so called "preservationists" and those who feel that the "wise use" of biodiversity is the best (and most realistic) road to its protection. Much of the work of big conservation groups in the North has been based on the propaganda of wise use, and has cited the manifold industrial potential of tropical biological diversity to justify the protection of these ecosystems (Ehrenfeld, 1990). Programmes seeking to secure increased marketing of rainforest products in the North are now being designed and implemented; promoted by conferences such as "The Rainforest Harvest" which took place in London in April 1990.

But the school of wise use has largely failed to give a sound ecological definition of exactly how much use is wise and at which level of exploitation the future of an ecosystem is imperiled (Ehrenfeld, 1990). Although any intensification in the use of biological resources bears the danger of overexploitation, the BCSP pays little attention to this problem. In principle, the assessment of the potential of certain plant or animal species to provide for a new market must include careful ecological auditing in every case. Otherwise, conservation through utilization will turn out to be a dangerous and destructive strategy.

It is not enough to speak about "sustainable use" in this context. The debate on "sustainable" levels of timber extraction in tropical rain-forests shows clearly how ambiguous this phrase has become. In many cases, all it describes is outright ecological destruction (Colchester, 1990). A biodiversity action program must provide against such forms of misuse from the beginning. At this stage, however, the BCSP is not taking up the challenge of defining ecological acceptable levels of harvest or designing mechanisms to implement such restrictions.

Not all of biodiversity can be translated into biological resources. Commercial uses will not be found for many species, even in the distant future. Indirectly, countless "useless" species benefit humanities through their role in the web of ecological interdependencies. The BCSP touches on these questions, but it is in danger of losing sight of the fact that the "bottom line" must always be defined in ecological terms.[2]

Who Will Use It?

The BCSP characterizes "using biodiversity" as "converting the un-known or little-appreciated wealth of nature into biological resources that contribute to human well-being, local community self-reliance, and environmental protection. This requires research and development efforts to create new uses of biological resources..." (Gift to the Future, 1989).

There are several misleading assumptions in this statement. Most communities that depend on intact nature are well aware of the importance of conserving natural diversity. In fact, such communities are far superior to modern industrial societies in terms of their relationship with nature, which is based on respect and a sense of community, instead of just viewing it as 'resources'.

This traditional perception of nature is well expressed in the following statement by a village elder from northern Thailand:

> "The forest is the source of the streams we use for farming and drinking water. We also rely on the forest when we build houses or make farming tools. We preserve the forest so that every year, when we need wood to repair the Muang Faai (traditional irrigation system), there is a place to find it. Every day we need to go into the forest to collect bits of dry wood to use as firewood. We use the forest as grazing grounds for our cows and buffalos and we collect bamboo shoots, mushrooms, fruits and vegetables from the forest to eat. When we are sick, we depend on herbal medicines from the forest. We repair our roofs with materials from the forest. There we find many of the things we need to make a living. The forest has bestowed kindness on us and we respect it in return" (in Permpongsacharoen, 1990).

The Programme's commitment to developing new uses for biodiversity rests on the assumption that by adding value to these 'resources', loss of biodiversity can be controlled. WRI's Research Programme (1989) states:

> "A revolution in our approach to the use of biodiversity is needed that will markedly increase the flow of benefits to people in the form of new jobs, local industry, substitution for imports, and new exports, and in so doing increase the rationale for maintaining and studying biodiversity".

In the face of existing uses by rural communities, the Programme's call for a revolution in "increasing biodiversity's utility" seems woefully misplaced. Where local communities have not already been displaced or dispossessed, they are often involved in intense and widespread resistance against the destruction of their resource base by outside economic forces. New export products cannot compensate for the destruction of existing subsistence economies. There is not so much a need to create new jobs as to preserve existing livelihoods.

Without a framework that addresses local rights, recognises traditional knowledge and supports grassroots conservation, the emphasis on 'new' uses for biodiversity could actually lessen the ability of local people to protect their lands and livelihoods. The BCSP should set as its goal the documentation of existing uses of biodiversity, helping to legitimize these uses and making them the basis for policy decisions on the use of rainforests.

There is, of course, a place in programmes for the conservation of biological diversity for developing new methods of utilization. But close attention must be paid to the fundamental issues of 'who will use it, and how will they use it'. The rubber tappers in the Amazon are often cited as a model for a sustainable and ecologically benign way of utilizing rainforest ecosystems for a wider market. This is so because their movement has a long term interest in the conservation of their forests.

In contrast, the first commercial rubber tappers were outsiders, who organized rapacious expeditions into the Amazon that practically annihilated the populations of the original rubber tree *Castilloa elastica*, and destroyed the indian communities that lived in these areas. The rubber boom led to the enslavement and annihilation of tens of thousands of Indians in Western Amazonia at the hands of the notorious Arana rubber company. Non-Indian rubber tappers suffered generations of debt slavery at the hands of the rubber barons.

When Hocker and Wickham brought seed material of the rubber tree *Hevea brasiliensis* to Asia, they laid the ground for the destruction of huge areas of forest in countries like Malaysia through the creation of rubber plantations. History teaches us that whenever large sums of money have been made from forests, without traditional control, local peoples have been overwhelmed and their forests pillaged. Only securing their rights and empowering them with control over their resources will check this process.

The Global Dimension

The work of the BCSP and many conservationists, especially in the North, is based on the idea that biodiversity is a global heritage. We share this view, but feel that it must be accompanied by a clear analysis of the role of consumer societies in the depletion of biological diversity. In our assessment the BCSP is inadequate in its approach, since it does not lay open, or sufficiently address, the global linkages in the destruction of nature in the Third World. As pointed out earlier in this essay, the massive destruction of rainforest ecosystems over the last decades can be directly traced to the resource hunger of industrialized countries. In a war of exploitation, huge areas of rainforests have been destroyed to win tropical timber, iron ore, bananas, soya, coffee and many more goods for the markets of the industrialized world.

These patterns of exploitation have rarely been adequately documented. A study by IUCN Holland (1988) found that the Netherlands is responsible for the degradation of an area in the tropics, as large as its own territory, to provide products such as soya and tapioca for its agricultural production. In the Brazilian state of Sao Paulo, 18 million hectares of rainforest, an area twice as big as Germany's forests, have been cleared to produce coffee for the international market (Griesshammer and Burg, 1989). Europe's steel industries are thriving on cheap iron ore imported at concessional rates from Brazil's Grande Carajas programme. Millions of hectares of forests and several Indian cultures have been decimated through this programme (Treece, 1987). The Federal Republic of Germany degrades approximately 200,000 hectares of rainforest each year just through its consumption of tropical timber (ARA, 1989). All these forms of exploitation invariably go hand in hand with the dispossession of local peoples as more and more areas come under the control of corrupt national elites, powerful companies and development planners.

The real challenge for a global programme to protect biodiversity is to examine these linkages. A global programme is necessary precisely because environmental problems in the Third World (and elsewhere) cannot be solved in a national context. Solutions are called for that will require to structural ecological adjustment within the affluent societies of the world. In our view, an integral part of the BCSP should be the drawing up of "National Conservation Strategies" for First World countries which addresses their impact on the global biosphere. At the same time, those Third World countries that have supported our wasteful lifestyles need to be offered technical and financial support to change back to sustainable patterns of livelihood and economy.

Of course, internal factors are responsible for a considerable portion of the loss of biodiversity taking place in the Third World. One of the greatest stumbling blocks to the conservation of tropical biodiversity is the control wielded by the ruling parties and elites of tropical countries who profit from the destruction of nature, and have no interest in giving up their present activities. Even so, it is the structural conditions set by a world economic order governed by the priorities of the North that limit the space in which Third World governments can make decisions over the direction of their development. In addition, many Third World elites and their despotic regimes are directly propped up by funds and political support from the developed world.

Even amongst progressive non-governmental organisations (NGOs), the importance of changing the patterns of exploitation of Northern governments, banks, businesses and consumers is often neglected. This is seen clearly in the recent discussions on the reformation of the Tropical Forestry Action Plan (TFAP). The solutions presented by the NGO community so far, do not include a plan for the North. The focus is still on the tropical countries themselves, as though a solution to deforestation can be found through funding a different kind of project, without attending to the international economic forces that generate and perpetuate the problem.

The South American agroecologist, Miguel Altieri (1989), characterized the bias in many Northern conservation initiatives in the following way:

> "The terminology associated with this view (sustainability, global conservation etc.) quickly permeated academic and development circles and led to acceptance of arguments and policies that favor securing the ecological resources and services of the ecosystems of the South for continual subsidy of the affluent, wasteful, and at times selfish development of the North. Thus the South is asked to engage in urgent and strict conservation, but the North is not asked to drastically reduce its level of consumerism and waste".

A "global" biodiversity program must state clearly and forcibly that the concept of a global heritage works both ways. Just as the protection of rainforests is a matter of global responsibility, so is the preservation of the North Sea or temperate forests. Generating an international movement which puts pressure on industrialized countries to protect their own endangered nature, and end their unsustainable use of resources, is an absolute necessity. This approach can become a powerful tool for linking up conservation movements from different parts of the globe.

The Roots of Change

The enormous potential that exists in the North for campaigns and actions in solidarity with social and environmental issues in the South is evident in the form of world-wide campaigns to protect rainforests and rainforest cultures by reducing the use of tropical timber. As a result of local education work and activism, over two thousand local councils throughout western Europe have passed resolutions to discontinue using tropical timber. Several European governments are in the process of banning the import of tropical timber unless its extraction conforms to strict ecological and social conditions. In other countries such as Australia, unions and timber user groups have stopped using tropical timber until such time as its destructive extraction from primary forests is stopped.

In our view, the ecological education of citizens in the North must play a central role in all strategies to halt the loss of biological diversity. The cliché of "think globally, act locally" can only be turned into practical politics if consumers are educated on the effects that their lifestyles are having on the nature and cultures of other regions. Yet, information on the political and economic causes of and solutions to the biodiversity crisis are not being provided to the public by most of the major conservation organisations or environmental education centres. Perhaps the greatest challenge for the environmental movement today is to find ways that the general concern of "the masses" for the protection of the environment can be translated into political action.

The BCSP seeks to support conservation from the public and private sectors without paying enough attention to how change towards environmental protection, both immediate and structural, comes about.

Experience shows that governments are moved to protect and conserve nature only when people push. Unless the Program makes the links with the grassroots, with the peoples and organisations that are out there resisting nature destruction, generating and forcing political change, it will fail.

The most important contribution that the BCSP could make to the global movement for the conservation of biological diversity is helping to link up grassroots initiatives. It could provide a framework for NGOs to make contact and work together to prepare common initiatives and come up with international demands and campaigns. One example of the kind of South-North cooperation that we envision are the climate pacts being initiated between cities and local councils in Western Europe and indigenous communities in Latin America. Cities such as Berlin, Frankfurt, Salzburg and Graz have committed themselves to drastically reducing their CO_2 output and at the same time committed funds towards the support of communities in the South such as the rubber tappers and Indians, who are actively protecting the rainforests.

At present, there are immense gaps and splits in understanding and perspective between the different groups working for the protection of nature. There is an urgent need to find a common language that can bridge the gap between scientific conservation initiatives on the one hand, and socially oriented NGOs working for nature protection on the other. Given the global interlinkages of the biodiversity crisis, no single approach to conservation can be effective. The success of the BCSP and other international conservation initiatives will depend on the degree to which they are able to bring together the different and often disparate parts of the conservation movement. The challenge stands for conservationists around the world to explore their stands and differences, learn from each other, and generate global initiatives that link local people's struggles, the spectrum of NGOs, scientific expertise, and where appropriate, various levels of government and their international agencies.

Footnote [1]. *Congo and Ghana with 75%, Liberia with 65%, Zaire with 63%, Papua New Guinea with 60%, Indonesia with 68%, Philippines with 62% and Thailand with 51% are all countries where wild vertebrates play an essential role in providing the population with protein, not to mention the plant food resources of the rainforests.*

Footnote [2]. *The BCSP has recently refined its guiding motto to: "Understand it, Defend it, and Use it Wisely and Equitably", and has incorporated some of the comments and concerns of the authors.*

References

Altieri, M., 1989. Rethinking Crop Genetic Resource Conservation: A View From the South, *Conservation Biology* Vol.3, Nr.1.

ARA/infoe, 1989. *Memorandum on the Responsibility of the Federal Republic of Germany for the Destruction of Tropical Rainforest and Her Obligation to Take Action for Their Protection*, P.O. Box 531 D-4800 Bielefeld, Germany.

ARA, 1989. *Schutz der Tropischen Regenwald* – Stelltlgnahme der Arbeitsgemeinschaft Regenwald und Artenschutz für die Enquete-Kommission des Deutschen Bundestag, Bielefeld.

Bank Information Centre, 1990. *Financing Ecological Destruction*, Washington D.C.

Calcagnotto, G., 1984. *Fallstudie: Der Sojaanbau in Brasilien*, epd Materialien 5/84, Germany.

Caufield, C., 1984. *In the Rainforest*, University of Chicago Press.

Chin, S.C., Devaraj, J., and Jin, K.K., 1989. *Logging Against the Natives of Sarawak*, INSAN, Selangor, Malaysia.

Colchester, M., 1989. *Pirates, Squatters and Poachers: The Political Ecology of Dispossession of the Native People of Sarawak*, Survival International, London.

Colchester, M., 1990. The International Tropical Timber Organisation: Kill or Cure for the Rainforests? in *The Ecologist* 20 (5).

Conservation International, 1990. *New Monkey Species Discovered in Brazil's Atlantic Forest,* Press release 20 June.

Daniel, J., and A. Kulasingham, 1974. Problems Arising from Large-scale Forest Clearing for Agricultural Use, *Malaysian Forester* 37.

Dosso, H., Guillaumt, J. and Hadley, M., 1981. The Tai Project: Land Use Problems in a Tropical Forest, *AMBIO* 10.

Ehrenfeld, D., 1990. "Conservation Biology: Wissenschaft fr das Leben" in ARA (ed), *Naturerbe Regenwald,* Focus Verlag Giessen.

Ehrlich, P., and Ehrlich, A., 1981. *Extinction. The Causes and Consequences of the Disappearance of Species,* Random House, New York.

Fajer, E., 1989. "How Enriched Carbon Dioxide Environments May Alter Biotic Systems Even in the Absence of Climatic Changes," *Conservation Biology* Vol.3, Nr.3.

Frankfurter Rundschau 21.11.88. *Auf der Flucht vor der Natur.*

Goldemberg, J., 1989. "Amazonia: Facts, Problems and Solutions," Speech at the International Symposium on the Amazon, July 31–August 2 in Sao Paulo.

Griesshammer, R., and Burg, C., 1989. *Wen die Banane krumm macht,* Reinbek, Germany.

Hong, E., 1987. *The Natives of Sarawak,* Third World Network, Penang, Malaysia.

IIASA in Glocke 11.4.90. Wissenschaftler warnen vor Verfall der Wälder Europas.

IUCN and UNEP, 1986. *Review of the Protected Areas System in the Afrotropical Realm,* Gland, Switzerland.

IUCN and UNEP, 1986. *Review of the Protected Areas System in the Indo-Malayan Realm,* Gland, Switzerland.

IUCN Netherlands, 1988. *The Netherlands and the World Ecology,* Amsterdam.

Jacobs, M., 1982. "The Study of Minor Forest Products," in *Flora Malesiana Bulletin 35.*

Janzen, D., 1986. "The External Threat," in Soulé, M. (ed.): *Conservation Biology – The Science of Scarcity and Diversity,* Sinauer Associates, Massachusetts.

Knoll, A., 1984. "Patterns of Extinctions in the Fossil Record of Vascular Plants" in Nitecki,M. (ed.): *Extinctions,* Chicago Press, Chicago.

LÖLF, 1986. *Rote Liste der in Nordrhein*-Westfalen gefährdeten Pflanzenarten, 2. Fassung, Germany.

Lutzenberger, J. 1987. "Who is destroying the Amazon Rainforest?" in *The Ecologist* 17 (4/5).

Lutzenberger, J., 1989. "Rainforests – Their Survival and Ours" in *Link Up* March.

McNeely, J., Miller, K.., Reid, W., Mittermeier, R. and Werner, T., 1990. *Conserving the World's Biological Diversity,* IUCN, WRI, CI, WWF-US and World Bank, Washington D.C.

Myers, N., 1985. *The Primary Source,* W.W. Norton, New York.

Myers, N., 1986. "Tropical Deforestation and a Mega-Extinction Spasm" in Soulé, M. (ed.): *Conservation Biology – The Science of Scarcity and Diversity,* Sinauer Associates Massachusetts.

Myers, N., 1988. *Natural Resource Systems and Human Exploitation Systems: Physiobiotic and Ecological Linkages,* Environment Department, World Bank.

Myers, N., 1989. *Deforestation Rates in Tropical Forests and their Climatic Implications,* Friends of the Earth, London.

Naess, A., 1986. "Intrinsic Value; Will the Defenders of Nature Please Rise" in Soulé, M. (ed.): *Conservation Biology – The Science of Scarcity and Diversity,* Sinauer Associates, Massachusetts.

Nectoux, F. and Y. Kuroda,1989. *Timber of the South Seas: An Analysis of Japan's Tropical Timber Trade and its Environmental Impact,.* WWF International, Gland Switzerland.

Ngau, H., 1989. Acceptance Speech for the Goldman Environmental Prize in *Utusan Konsumer,* CAP, Penang, Malaysia.

Permpongsacharoen, W., 1990. "Widerstand gegen Waldzerstörung-Lektionen aus den Dörfern Thailands," in ARA (ed.): *Naturerbe Regenwald,* Focus Verlag, Giessen, Germany.

Prance, G., Bale, W., Boom, B. and Carneiro, R., 1987. "Quantitative Ethnobotany and the Case for Conservation in Amazonia," in *Conservation Biology* 1(4).

Raup, D., 1988. "Diversity Crisis in the Geological Past" in Wilson, E.O. (ed.): *Biodiversity,* National Academy Press, Washington.

Raven, P., 1988. "Our Diminishing Tropical Forests" in E.O. Wilson (ed.): *Biodiversity,* National Academy Press, Washington.

Raven, P., 1989. cited in "The Death of Birth," *TIME* Vol.133 Nr.1.

Secrett, C., 1986. *Rainforest – Protecting the Planet's Richest Resource,* Friends of the Earth.

Simberloff, D., 1986. "Are We on the Verge of a Mass Extinction in Tropical Rain Forests?" in Eliott, D. (ed.): *Dynamics of Extinction,* New York.

Soulé, M. and Wilcox, B., 1980. "Conservation Biology – its Scope and its Challenge" in *Conservation Biology: an Evolutionary-Ecological Perspective,* Sinauer Associates, Massachusetts.

Treece, D., 1987. *Bound in Misery and Iron,* Survival International, London.

Vitousek, P., Ehrlich, P., Ehrlich, A. and Matson, P., 1986. "Human Appropriation of the Products of Photosynthesis." *Bioscience 36(6).*

Wilson, E.O., 1988. "The Current State of Biological Diversity" in Wilson, E.O. (ed.): *Biodiversity,* National Academy Press, Washington.

Wilson, E.O., 1990. "Bedrohung des Artenreichtums" in *SPEKTRUM,* Sonderheft 9.

World Bank, 1988. *Annual Report 1988,* Washington D.C.

WRI, IUCN and UNEP, 1989. *Gift to the Future; Program Description,* Washington D.C.

WRI, IUCN and UNEP, 1989. *Gift to the Future; Research Topics,* Washington D.C.

Biodiversity, Biotechnology and Profits.

Vandana Shiva

Biotechnology and the production of uniformity

As biology takes the centre stage of industrial civilization with the emergence of the new biotechnologies, biodiversity conservation also moves centre stage for the green movement.

There are, however, not one but many strands in the conservation of biodiversity. The first is inspired by the deep green ethos of the democracy of all life. It is based primarily on the ethical ground that all life forms have value in themselves, independent of the value man puts on them.

The second strand shares this ethical concern and to it adds the dimension of justice and equity. Arising largely from the peasant and tribal movements in the South, biodiversity conservation takes on the significance of a struggle for self-reliance and decentralization. Diversity in this perspective needs to be incorporated into the logic of production processes in forestry and agriculture because diversity protects the livelihood niches of different communities.

The third approach is related to the emergence of the new biotechnologies which have transformed the richness of this planet into a strategic raw material for industrial production of food,pharmaceutical, fibres, energy, etc. Biodiversity conservation then becomes merely conservation of 'raw material' rather than conservation of 'means of production' of life itself.

The schizophrenic approach to production and conservation with 'production' being based on uniformity, and 'conservation' desperately attempting to preserve diversity as raw material against the force of expanding monocultures, guarantees that neither biodiversity nor peoples' access to it will be protected. They can only be protected by making diversity the **basis** and **foundation** of production and economic activity, not merely an 'input'. This challenge of **conserving biodiversity in production** is being systematically missed by all programmes and projects that deal with the issue of biodiversity. Thus, for example, *World Resources 1990-1991* published by World Resources Institute (WRI) in collaboration with the United Nations Environment Programme (UNEP) and United Nations Development Programme (UNDP) has a chapter on "Food and Agriculture", another on "Forests and Rangeland", and a third on "Wildlife and Habitat". While all three chapters should deal with biological diversity, it is only the chapter on wildlife which has a section on biodiversity conservation. This myopic approach to the conservation of biodiversity which accepts, and indeed encourages the destruction of diversity in spheres of production such as agriculture, forestry, animal husbandry and fisheries is also reflected in most global discussions on biodiversity. On one hand this narrows down the challenge of biodiversity conservation into a sub-field of wildlife management. On the other hand, it avoids addressing the major ecological, political and economic issues related to the use and control over biological resources, and the impact on biodiversity of the new technologies.

There is a prevalent misconception that biotechnology development will automatically lead to biodiversity conservation. The main problem with viewing biotechnology as a miracle solution to the biodiversity crisis is related to the fact that biotechnologies are, in essence, technologies for the breeding of uniformity in plants and animals. Biotechnology corporations do talk of contributing to genetic biodiversity. As John Deusing of Ciba-Geigy states,

"Patent protection will serve to stimulate the development of competing and diverse genetic solutions with access to these diverse solutions ensured by free market forces at work in biotechnology, ecology and seed industries" (1).

However, the diversity of corporate strategies and the diversity of lifeforms on this planet are not the same thing, and corporate competition can hardly be treated as a substitute for nature's evolution in the creation of genetic diversity.

Corporate strategies and products can lead to diversification of commodities, they cannot enrich nature's diversity. This confusion between commodity diversification and biodiversity conservation finds its parallel in raw material diversification. Although breeders draw genetic materials from many places as raw material input, the seed commodity that is sold back to farmers is characterised by uniformity. Uniformity and monopolistic seed supplies go hand in hand. When this monopolising control is achieved through the molecular mind, destruction of diversity becomes more accelerated. As Jack Kloppenburg has warned,

"Though the capacity to move genetic material **between** species is a means for introducing additional variation, it is also a means for engineering genetic uniformity **across** species" (2).

Productivity and Diversity

According to the dominant paradigm of production, diversity goes against productivity, which creates an imperative for uniformity and monocultures. This has generated the paradoxical situation in which modern plant improvement has been based on the destruction of the biodiversity which it uses as raw material. The irony of plant and animal breeding is that it destroys the very building blocks on which the technology depends. Forestry development schemes introduce monocultures of industrial species such as eucalyptus, and push into extinction the diversity of local species which fulfills local needs. Agricultural modernisation schemes introduce new and uniform crops into farmers' fields and destroy the diversity of local varieties. In the words of Professor Garrison Wilkes of the University of Massachussets, this is analogous to taking stones from the foundation of a building in order to repair the roof. This strategy of basing productivity increase on the destruction of diversity is dangerous and unnecessary.

Monocultures are ecologically unstable, and that reason alone should be enough to NOT view them as essential to production. Monocultures invite diseases and pests, as was experienced in 1970-71 in the U.S. with the corn blight epidemic which laid waste 15 percent of the nation's crop, because of genetic uniformity. 80 percent of the hybrid corn in U.S. in 1970 was planted with corn which contained **T. cytoplasm** which made the corn plants vulnerable to the corn blight fungus, *H. maydis.* It had been used by plant breeders and seed companies to foster quick and profitable production of high-yielding, hybrid corn seed, but the introduced cytoplasm was vulnerable to corn blight.

According to the 1972 National Academy of Sciences study on *Genetic Vulnerability of Major Crops,* 'The corn crop fell victim to the epidemic because of a quirk in the technology that had designed the corn plants of America, until, in one sense, they had become as alike as identical twins. Whatever made one plant susceptible make them all susceptible" (3).

The technology for breeding high yield varieties is therefore a technology which breeds uniformity and threatens a collapse in yields. What America experienced with corn, Asia has experienced with the Green Revolution rice varieties.

Centralised research and genetic uniformity go hand in hand in agriculture. It was therefore inevitable that CIMMYT (Centre for Wheat and Maize improvement) in Mexico became an instrument for the destruction of genetic diversity of maize, while IRRI (International Rice Research Institute) in the Philippines became an instrument for the destruction of genetic diversity of rice in Asia.

Wherever the new IRRI rice varieties were introduced in Asia, they proved to be susceptible to diseases and pests. For example, IR-8 was attacked by bacterial blight in South-east Asia in 1968 and 1969. In 1970 and 1971, it was destroyed by the tungro virus. In 1975, half a million acres under the new rice varieties in Indonesia were destroyed by pests. In 1977, IR-36 which was supposed to be resistant to eight major known diseases and pests including bacterial blight and tungro virus, was developed. However, it was attacked by two new viruses called 'ragged stunt' and 'wilted stunt'. In the State of Punjab in India, insects and pests which were considered insignificant before the Green Revolution have now become major problems. At present, rice cultivation in Punjab is vulnerable to about 40 insects and 12 diseases (4).

In agriculture and forestry, in fisheries and animal husbandry, production is being incessantly pushed in the direction of diversity destruction. Production based on uniformity thus becomes the primary threat to biodiversity conservation and to sustainability.

Not till diversity is made the logic of production can diversity be conserved. If production continues to be based on the logic of uniformity and homogenisation, uniformity will continue to displace diversity. 'Improvement' from the corporate viewpoint, or from the viewpoint of western agricultural or forestry research, is often a loss for the Third World, especially for the poor in the Third World. There is therefore no inevitability that production should act against diversity. Uniformity as a pattern of production becomes inevitable only in a context of control and profitability.

The spread of monocultures of 'fast-growing species' in forestry and 'high-yielding varieties' in agriculture has been justified on grounds of increased productivity. All technological transformation of biodiversity is justified in the language of 'improvement' and increase of 'economic value'. However, 'improvement' and 'value' are not neutral terms. They are contextual and value-laden. Improvement of tree species means one thing for a paper corporation which needs pulping wood and an entirely diffferent thing for a peasant who needs fodder and green manure. Improvement of crop species means one thing for a processing industry and something totally different for a self-provisioning farmer.

However, categories of 'yield', 'productivity', and 'improvement' which have emerged from the corporate viewpoint have been treated as universal and value mentral. Thus, all tree-planting programmes financed by international institutions in recent years and encouraged by the Tropical Forestry Action Plan (TFAP) have spread eucalyptus monocultures across Asia, Africa and Latin America. The argument used most frequently in favour of the spread of eucalyptus monocultures is that it is fast growing. However, the only fast growth that eucalyptus contributes to is pulp wood. It is not fast growing in terms of yield of wood for other purposes. In terms of yields of non-woody biomass for fodder, eucalyptus has zero yield since its leaves are not eaten by cattle (5). Given that the industrial sector does not benefit from the diversity of species and uses of trees, forestry programmes deliberately destroy diversity for increasing yields of industrial raw material. Thus, in spite of high diversity and high productivity of tropical forests which yield up to 300 tons of wood per hectare, as compared to 150 tons per hectare for temperate forests, naturally diverse tropical forests are considered 'unproductive'. Referring to the diversity and large biomass of tropical forests, a forestry expert has stated 'that from a standpoint of industrial material supply, this is relatively unimportant. The important question is how much of this biomass represents trees and parts of trees of preferred species that can be profitably marketed today. By current utilisation standards, most of the trees in these tropical forests are, from an industrial materials standpoint, clearly weeds' (6).

Viewing diversity as weeds leads to the extinction of that diversity which has high ecological and social values even when it does not profit industry. The pattern of the destruction of diversity has been the same in forestry and agriculture.

Plant improvement in agriculture has been based on the 'enhancement' of the yield of desired product at the expense of unwanted plant parts. The 'desired' product is however not the same for agribusinesses and Third World peasants. Which parts of a farming system will be treated as 'unwanted' depends on what class and gender one is. What is unwanted for agribusiness may be wanted by the poor, and by squeezing out those aspects of biodiversity, agriculture 'development' fosters poverty and ecological decline.

In India, the 'high-yielding' strategy of the Green Revolution squeezed out pulses and oilseeds which were essential for nutrition and soil fertility. The monocultures of the dwarf varieties of wheat and rice also squeezed out the straw which was essential for fodder and fertilising the soil. The yields were 'high' from the viewpoint of centralised control of food-grain trade, but not in the context of diversity of species and products at the level of the farm and the farmer.

Figure 1. 'Productivity': Monocultures vs. Diverse Systems

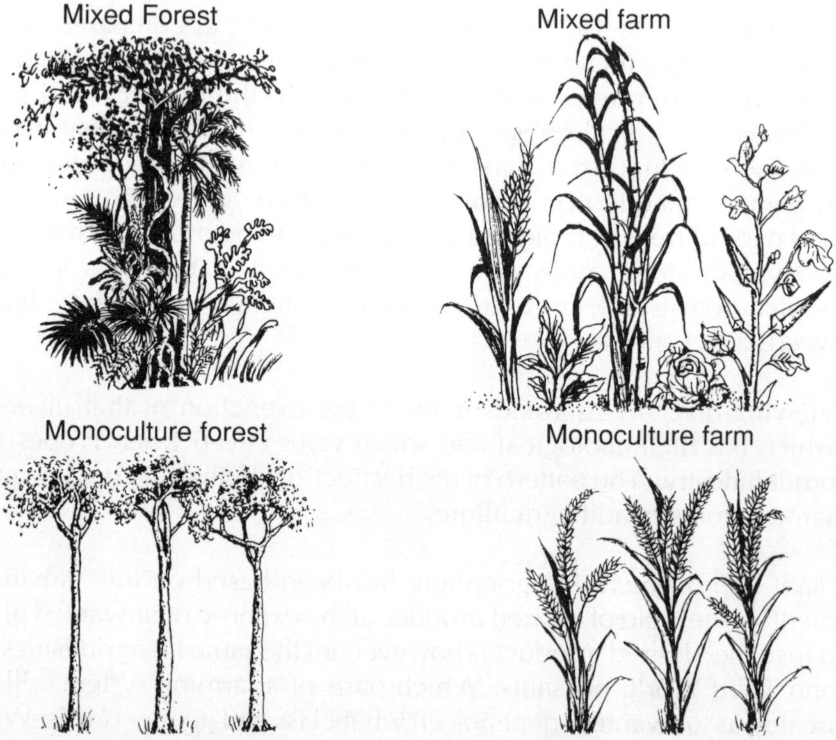

Overall productivity and sustainability is much higher in mixed systems of farming and forestry which produce diverse outputs.

Productivity of monocultures is low in the context of diverse outputs and needs. It is high only in the restricted context of output of "part of a part" of the forest and farm biomass. E.g. "High yield" plantations pick one tree species among thousands, for yields of one part of the tree (e.g. pulpwood).

"High yield" green revolution cropping patterns pick one crop among hundreds, e.g. wheat for yields of one part of the wheat plant (only grain).

These high partial yields do not translate into high total (including diverse) yields.

Productivity is therefore different depending on whether it is measured in a framework of diversity or uniformity.

Biodiversity: 'Raw Material' or 'Means of Production'

For peasants and forest-dwellers, biodiversity has been the source of sustenance for basic needs such as food, clothing, shelter, and medicine. To use Jack Kloppenburg's analogy of the seed: it is both a 'means of production' as well as a 'product'. Whether it is the tribal doing 'shifting cultivation' or the peasant practising settled agriculture, in planting each year's crop, the farmers also reproduce the necessary part of their means of production. The seed thus represents capital with a simple biological obstacle; given the appropriate conditions it reproduces itself and multiplies. Modern plant-breeding has primarily been an attempt to remove this biological obstacle, and the new biotechnologies are the latest tools for transforming what is simultaneously a 'means of production' and 'product' into mere 'raw material'.

The cycle of regeneration of biodiversity is therefore replaced by a linear flow of free germplasm from farms and forests into corporate labs and research stations, and the flow of modified uniform products as priced commodities from corporations to farms and forests. Diversity is destroyed by transforming it into mere raw material for industrial production based on uniformity.

This change in the nature of biodiversity is exemplified in the case of seeds. New technologies treat self-regenerative seed as 'primitive, raw' germplasm, and the seed that is inert without inputs and non-reproducible as a 'finished product'. The whole is rendered partial, and the partial is rendered whole. The commoditised seed is ecologically incomplete and ruptured at two levels:

(1) It does not **reproduce** itself, while by definition, seed is a generative

resource. Through technological transformation, biodiversity is transformed from a renewable into a non-renewable resource.

(2) It does not **produce** by itself. It needs the help of inputs to produce. As the seed and chemical companies merge, the dependence on inputs will increase, not decrease. Whether a chemical is added externally or internally, as in the new biotechnologies, it remains an external input in the ecological cycle of the reproduction of seed.

It is this shift from the ecological processes of reproduction to the technologies processes of production that underlies both the problem of dispossession of farmers and tribals and the problem of erosion of biodiversity.

Uniformity becomes the primary threat to biodiversity conservation, even though in the convoluted political economy of the market, it is cited as the reason for conservation. 'The exploitation of genetic diversity for crop improvement should be the ultimate objective of genetic resources exploration and conservation', it is argued. The arbitrary inequality created in the status of germplasm creates an arbitrary separation between production and conservation. Some peoples' germplasm becomes a finished commodity, a 'product'. Other peoples' germplasm becomes mere 'raw' material for that product. The manufacture of the 'product' in corporate labs is regarded as production. The reproduction of the 'raw' material by nature and Third World farmers and forest dwellers is mere conservation. The 'value added' in one domain is built on the 'value robbed' in another domain. Biotechnology development thus translates into biodiversity erosion and poverty creation.

The challenge of the '90's is based on our getting rid of false notions ot obsolescence and productivity which legitimise the extinction of large parts of nature and society. The push for homogenisation and uniformity comes both from the transnational corporate sector, which has to create uniformity to control markets. It also comes from the nature of modern research systems which have grown in response to the market. Since most biotechnology research is dictated by TNCs, the solutions sought out must have a global and homogeneous character. TNCs do not tend to work for small market niches, but aim at large market shares. In addition, researchers prefer tasks that can be simplified enough to be tackled systematically, and that produce stable and widely applicable outcomes. Diversity goes against the standardisation of scientific research.

The main challenge to biodiversity conservation is the removal of reductionist blinkers that veil the social construction of 'growth and productivity' which is achieved by excluding plants and parts of plants as 'unwanted' and 'waste'.

However, more production of partial outputs as measured on the market and in a monoculture is often less production when measured in the diversity of nature's economy or peoples' sustenance economy. In the context of diversity, increased production and improved productivity can be consonant with biodiversity conservation. In fact, it is often dependent on it.

All systems of sustainable agriculture and forestry, whether of the past or the future, work on the basis of the perennial principles of diversity and reciprocity. The two principles are not independent but interrelated. Diversity gives rise to the ecological space for give and take, for mutuality and reciprocity. Destruction of diversity is linked to the creation of monocultures. With the creation of the monocultures, the self-regulation and decentred organisation of diverse systems gives way to external inputs and external, centralised control. Sustainability and diversity are ecologically linked because diversity offers the multiplicity of interactions which can heal ecological disturbance to the part of the system. Non-sustainability and uniformity means that a disturbance to one part is translated into a disturbance to all the other parts. Instead of being contained, ecological destabilisation tends to be amplified. Closely linked to the issue of diversity and uniformity is the issue of productivity. Higher yields and higher production have been the main push for the introduction of uniformity and the logic of the assembly line. The imperative of growth generates the imperative for monocultures. Yet this growth is, in large measure, a socially constructed, value laden category. It exists as a 'fact' by excluding and erasing the facts of diversity and production through diversity. Sustainability, diversity and justice through decentered self-organisation are therefore linked, as are unsustainability, uniformity, centralisation and justice.

Figure 2. Biodiversity: 'Means of production' or 'raw material'.

In traditional agriculture and forestry systems, biodiversity regenerates itself. It is both 'means of production'and 'product'. Thus, seed has been the source of grain as well as future source of seed.

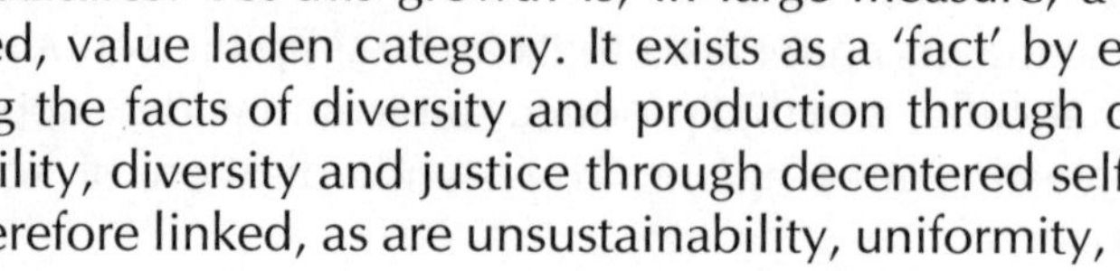

Industrialised agriculture and the seed industry, especially with the advent of the new biotechnologies, convert this seed into mere raw material for seed as commodity, patented and privatised.

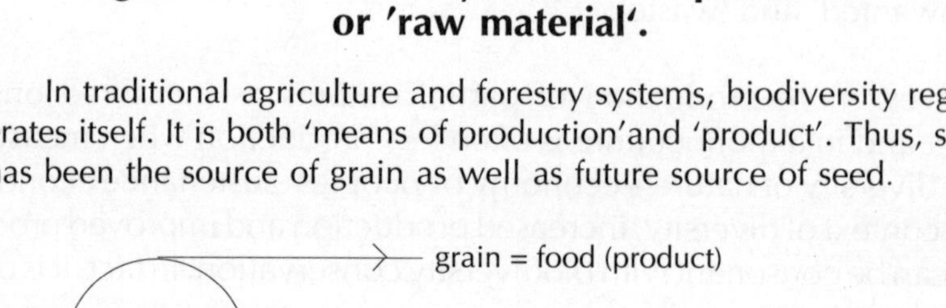

Patents, Intellectual Property and the Politics of Knowledge

Farmers' seeds are rendered incomplete and valueless by the process that makes corporate seeds the basis of wealth creation. The indigenous varieties or land races, evolved through both natural and human selection, and produced and used by Third World farmers world-wide are called 'primitive cultivars'. Those varieties created by modern plant breeders in international research centres or by transnational seed corporations are called 'advanced' or 'elite'. The tacit hierarchy in words like 'primitive' and 'elite' becomes an explicit one in the process of conflict. Thus, the North has always used Third World germplasm as a freely available resource and treated it as valueless. The advanced capitalist nations wish to retain free access to the developing world's storehouse of genetic diversity, while the South like to have the proprietory varieties of the North's industry declared a similarly 'public' good. The North, however, resists this democracy, based on the logic of the market. Williams, Executive Secretary of IBPGR has argued that **'it is not the original material which cash returns'**. A 1983 forum on Plant Breeding, sponsored by Pioneer Hi-Bred stated that:

'Some insists that since germplasm is a resource belonging to the public, such improved varieties would be supplied to farmers in the source country at either zero or low cost. This overlooks the fact that **"raw" germplasm only becomes valuable after considerable investment of time and money**, both in adapting exotic germplasm for use by applied plant breeders in incorporating the germplasm into varieties useful to farmers' (7).

The corporate perspective views as value only that which serves the market. However, all material processes also serve ecological needs and social needs, and these needs are undermined by the monopolising tendency of corporations.

The issue of patent protection for modified life forms raises a number of unresolved political questions about ownership and control of genetic resources. The problem is that in manipulating life forms you do not start from nothing, but from other life-forms which belong to others - may be through customary law. Secondly, genetic engineering biotechnology does not create new genes, it merely relocates genes already existing in organisms. In making genes the object of value through the patent systems, a dangerous shift is placed in the approach to genetic resources.

Putting value on the gene through patents makes biology stand on its head. Complex organisms which have evolved over millenia in nature, and through the contributions of Third World peasants, tribals and healers are reduced to their parts, and treated as mere inputs into genetic engineering. Patenting of genes thus leads to a devaluation of life-forms by reducing them to their constituents and allowing them to be repeatedly owned as private property. This reductionism and fragmentation might be convenient for commercial concerns but it violates the integrity of life as well as the common property rights of Third World peoples. On these false notions of genetic resources and their ownership through intellectual property rights are based the 'bio-battles' at FAO and the trade wars at GATT. Countries like US are using trade as a means of enforcing their patent laws and intellectual property rights on the sovereign nations of the Third World. US has accused countries of the Third World as engaging in 'unfair trading practice' if they fail to adopt US patent laws which allow monopoly rights in life forms. Yet it is the US which has engaged in unfair practices related to the use of Third World genetic resources. It has freely taken the biological diversity of the Third World to spin millions of dollars of profits, none of which have been shared with Third World countries, the original owners of the germplasm. A wild tomato variety (*Lycopersicon chomrelewskii*) taken from Peru in 1962 has contributed $8 million a year to the American tomato processing industry by increasing the content of soluble solids. Yet, none of these profits or benefits have been shared with Peru, the original source of the genetic material.

According to Prescott-Allen, wild varieties contributed US$340 million per year between 1976 and 1980 to the US farm economy. The total contribution of wild germplasm to the American economy has been US$66 billion, which is more than the total international debt of Mexico and Philippines combined. This wild material is 'owned' by Sovereign states and by local people (8).

Most Third World countries view genetic resources as common heritage. In most countries, animals and plants were excluded from the patent system until recently when the advent of biotechnologies changed concepts of ownership of life. With the new biotechnologies life can now be owned. The potential for gene manipulation reduces the organism to its genetic constituents. Centuries of innovation are totally devalued to give monopoly rights on life forms to those who manipulate genes with new technologies, placing their contribution over and above the intellectual contribution of generations of Third World farmers for over ten thousand years in the areas of conservation, breeding, domestication and development of plant and animal genetic resources. As Pat Mooney has argued, 'the argument that intellectual property is only recognisable when performed in laboratories with white lab coats is fundamentally a racist view of scientific development' (9).

Two biases are inherent in this argument. One, that the labour of Third World farmers has no value, while labour of western scientists adds value. Secondly, that value is a measure only in the market. However, it is recognised that, 'the total genetic change achieved by farmers over the millenia was far greater than that achieved by the last hundred or two years of more systematic science-based efforts'. Plant scientists are not the sole producers of utility in seed.

This utility of farmers' and tribals' seeds has high social and ecological value, even if it has no market value attached to it. The limits of the market system in assigning value can hardly be a reason for denying value to farmers' seeds and nature's seeds. It points more to the deficiency of the logic of the market than the status of the seed or the farmers' intellect.

There is no epistemological justification for treating some germplasm as valueless and common heritage and other germplasm as a valuable commodity and private property. This distinction is not based on the nature of the germplasm, but on the nature of political and economic power.

References

1. WRI, UNEP, UNDP, 1991. *World Resources 1990-91.*

2. Kloppenburg, J., 1988. *First the Seed,* Cambridge University Press, Cambridge, U.K.

3. Doyle, J., 1985. *Altered Harvest,* Viking.

4. Shiva, *V., 1991. Violence of the Green Revolution,* Third World Network, Penang.

5. Shiva, V. and J. Bandyopadhyay, 1987. *Ecological Audit of Eucalyptus Cultivation,* Research Foundation for Science and Ecology, Dehra Dunn, India.

6. Quoted by Vandana Shiva, *Forestry Crisis and Forestry Myths,* World Rainforest Movement, Penang.

7. Quoted by Jack Kloppenburg in *First the Seed.*

8. Iltis, H., 1986. *Serendipity in* "Exploration of Biodiversity: What Good are Weedy Tomatoes," in E.O. Wilson (ed) *Biodiversity,* National Academy Press.

9. Mooney, P., 1989. *'From Cabbages to Kings'. Intellectual Property vs. Intellectual Integrity,* Proceedings of ICDA Conference on Patents, Brussels.

The Impact of Biodiversity Conservation on Indigenous Peoples

Andrew Gray

The Biodiversity Crisis

Biological diversity is currently the most critical issue facing conservationists. The threat to "biodiversity" is so acute that it has become a global question. Several initiatives are under way to conserve the world's biodiversity and publications on the subject abound (Reid and Miller, 1989; McNeely *et al.*, 1990; WRI *et al.*, 1990). However, the analysis of the problem and the proposals suggested to solve the crisis repeat many previous deficiencies in environmental and resource management thinking.

The fundamental problem lies in the position of human beings in the analysis of the crises. This paper looks at the case of indigenous peoples in the context of biological diversity conservation discussion and how they risk major violations of their human rights if some of the policies now under discussions were to go ahead.

"Biodiversity is the variety of the world's organisms, including their genetic diversity and the assemblages they form" (Reid and Miller, 1989:3). This totality of genes, species and ecosystems provide the basis of life on earth. Estimates of the number of species in the world are more than 30 million.

The importance of biodiversity for the continuation of life rests on several factors. An ecosystem operates through the functional inter-relationship of species. When some of these species are destroyed the whole form of the system changes. This not only has consequences for the local environment but can have devastating effects on the inter-relationships between ecosystems.

Furthermore, biological diversity is important for encouraging alternative varieties of agricultural species which act as a protection against over-reliance on any particular variety. Similarly, it ensures the continuation of ecosystems as a whole and thereby the survival of as many different species as possible.

Certain areas of the world known as "Vavilov centres of genetic diversity" contain the greatest diversity of species. The peak of biological diversity is reached in tropical rainforests, coral reefs and island ecosystems. Nevertheless, the principle of inter-relationship between species holds for ecosystems throughout the world.

Most of the environmental discussion on biodiversity draws attention to the increasing devastation of the rainforest (Head & Heinzman, 1990). In 1989, Norman Myers estimated that 142,000 square kilometres of tropical forest were destroyed (Myers, 1989). Calculating the current rates of species extinction, Reid and Miller (1989:31) point to "perhaps 100 to 1,000 times what they would be in unperturbed nature". Taking a broader look on the basis of discussion in Wilson and Peter (1988), the overall figure could be as high as 30,000 times.

The Cultural Diversity Crisis

Although the rainforests of the world contain many different peoples, this paper will concentrate on the threats facing indigenous peoples. Indigenous peoples are the descendants of inhabitants of a country prior to its colonization, whose rights are still not fully recognised. There are several aspects to identifying indigenous peoples ranging from their relationship to the nation state to their own self-definition.

Indigenous peoples usually live in a particular territory and practise their own distinct social and cultural ways of life. Indigenous peoples are found throughout the world and number about 200 million people, about 4 percent of the global population (Burger, 1987:11). Although indigenous peoples have been fighting for their rights for hundreds of years, the last 25 years have seen the rise of an indigenous movement on an international level.

Through their lobbying, indigenous peoples' rights are gradually being recognised by the International Labour Organisation and the United Nations. Several countries have very strong constitutional and national legal provisions for recognising their rights to land and culture. However, in most parts of the world, indigenous peoples are at the receiving end of racial discrimination, territorial invasions, cultural ethnocide and, in some cases, genocide.

The indigenous peoples of the world have been increasingly demanding that the recognition of their fundamental right of self-determination is the only way to resolve the threats facing them. This means recognition and respect for the territorial and cultural rights of indigenous peoples, enabling them to take control of their own destiny.

The University of Bergen estimates that there are about 5000 different cultures in the world (G. Henriksen, pers. comm.). If we compare this figure with the number of national state cultures in the world, we see that indigenous peoples constitute 90-95 percent of the cultural diversity of the world. The greatest diversity of these cultures can be found in the rainforest areas of Central and South America, Africa, Southeast Asia and New Guinea.

The area of the highest biodiversity in the world are the rainforests which are the homelands of many indigenous peoples who are the original or prior inhabitants of the area. From this, one would assume that indigenous peoples would take pride of place in a world biodiversity plan. Unfortunately, this is not the case.

Biodiversity Conservation Initiatives

During the last three years, plans for a Biodiversity Conservation Strategy have been developed by the World Resource Institute (WRI) and the International Union for Conservation of Nature and Natural Resources (IUCN) with support from the World Bank and other institutions. Although it has undergone several modifications the plan has retained its essential elements. Based on the slogan "save it, study it, use it", the programme consists of a package including protected reserves, financing "debt-for-nature" swaps, bilateral conservation projects with governments and expanding the role of the private sector in harvesting forest products in a sustainable manner (WRI *et al.,* 1990).

Protection Zones

A major aspect of biodiversity conservation strategy plans take place in the field. This *"in situ"* conservation refers to protection zones in areas of high biological diversity. These areas, described as "natural ecosystems", will protect species with minimum of human interference. The "buffer zones" or "semi-natural" ecosystems can allow for some human disturbance as long as the "impact of humanity should not be greater than any other factor" (Reid and Miller, op. cit. :67).

However,the plans do not address the long history of conflict between indigenous peoples and those controlling national parks. These conflicts arise from various causes. A recurring pattern is that the indigenous peoples living in the areas designated as "natural ecosystem" are forcibly relocated or "encouraged" to move out to the buffer zones. There they are on the one hand encouraged to enter the market economy through using the resources extracted from the forest, while on the other hand they have to compete with colonists and other forest peoples already in the area. The result is invariably social unrest and poverty.

Documented examples demonstrate the problem. To satisfy the environmental goals of the government, the Ik of Uganda were forcibly relocated in order to establish the Kidepo Valley National Park (Turnbull, 1972). More recently in 1988, Maasai in Tanzania were removed from the village of Mkomazi to create a buffer zone for the Ngorongoro Park (IWGIA, 1989). Cases from Asia include the forest peoples of Sri Lanka, known as Vedda (Stegeborn,1985) and the dangers facing the villages which have found themselves placed with the tiger parks in West Bengal (IWGIA, 1986).

The parallel plans, such as debt-for-nature swaps have led to similar problems. The idea behind these swaps is to re-arrange the enormous national debts in return for conservation measures. Several countries have made such arrangements including Costa Rica, Ecuador and Bolivia. Bolivia agreed to spend $250,000 to protect the Chimenes forest, however the Chimenes Indians were not consulted and were encouraged to move to a "buffer" zone already being exploited by ranchers and loggers (Hecht and Cockburn, 1989). Although attempts are being made to improve the situation (Kent Redford, pers. comm.), the original planning was deficient to the extreme.

Extractive Reserves and Marketing

A principle idea behind the proposed Biodiversity Conservation Strategy, and indeed behind many forms of conservation nowadays is that conservation should be co-ordinated with the utilisation of its resource potential to generate profit. Profitability has become the key word for what is known as "green capitalism" (Elkington and Burke, 1989). In the context of biodiversity, this means making an evaluation of the species in the rainforest:

"Many actions that can be taken to stem the loss of biodiversity do provide short-term economic benefits - say, maintaining natural forests so that wild species can be harvested for food and medicines, and industrial products or establishing protected areas so that tourists will visit" (Reid & Miller, op. cit. :89).

Whereas this may be useful on a local scale, as a major facet of a biological diversity plan, aimed at international funding sources such as the World Bank, national governments and private business, it is more of an advertisement attracting investors to the resources of the rainforest than towards a conservation programme. This is exactly the criticism which has been raised against the Tropical Forestry Action Plan (Colchester & Lohmann, 1990):

"The Plan has been elaborated with almost no consultation with NGOs nor with community-based organisations, it paid little attention to the needs and rights of forest dwellers and seemed unduly focused on funding commercial forestry and wood-based industries, while failing to identify the real causes of deforestation" (*ibid*:2).

Nevertheless, resource managers working with environmental questions point to the need for indigenous and other forest peoples to gain an income to survive and suggest ways in which they can be tapped into the capitalist economy of the north. The strategy is thus to tie conservation into a sustainable means of development. This approach is also a strategy of some indigenous support organisations (Baker, 1989).

Unfortunately, there are many problems with this approach as it applies to indigenous peoples (Pearce, 1990). Even though many indigenous peoples want to exchange forest products and have been trading and bartering for centuries (Wolf, 1982), encouraging marketing of the rainforest without considering the issue from indigenous peoples' perspectives can have serious consequences.

Indigenous peoples have an economy which is controlled by their social relations and fits within their cultural framework. Trading and exchange would traditionally take place within this context with production decided and determined by the community.

If indigenous peoples enter the market economy on their own terms this can continue. However, once the demand from outside begins to determine production, this can quickly surpass the priorities of a subsistence and self-sufficient economy. The result is that the community ends up as an effective "wage-labourer" for the demands of consumers in the North.

In the past, this had terrible effects on the indigenous population. The rubber boom, the highways of Brazil, and the Peruvian Amazonian colonisation plans of the 1980s all had devastating effects on indigenous peoples by attracting them into a market economy over which they had no control. Indeed, the rubber boom, which resulted in the deaths of thousands of Amazonian Indians, was portrayed at the time as a form of sustainable development (Gray, 1990).

"Development" for the resource managers working on the environment is essentially a matter of economics. This counteracts the direction in which development discussions have been going in the last few years. Apart from concentrating on local initiatives, indigenous development projects strive to draw together "cultural" and "political" dimensions of development with sustainable approaches. The result is a "self-development" (Henriksen, 1989) which places the responsibility and control over projects in the hands of indigenous communities themselves.

Marketing for indigenous peoples is important but is not their main priority. Unless they receive respect for and recognition of their fundamental rights to control and determine their own lives and can live on their own territories without unnecessary interference, marketing becomes a luxury which they may well not be in a position to enjoy.

Indigenous peoples want to develop their own marketing strategies rather than have plans made from outside imposed on them. As an indigenous leader once told me "we prefer to make our own mistakes rather than having the mistakes of others thrust down our throats".

Intellectual Property Rights and Indigenous Knowledge

Indigenous peoples have a vast wealth of knowledge about their environment built up over centuries. Not only does this knowledge include information about different species of animals and plants, their behaviour and uses, but also information about the way in which aspects of the universe inter-relate.

For indigenous peoples, knowledge of the environment depends on contacts with the invisible spirit world which plays its own crucial part in ensuring the reproduction of society, culture and the environment. Whereas non-indigenous environmentalists deal exclusively with the relationship between human beings and "nature", indigenous peoples have a different way of conceptualising this. For them, knowledge is both material and spiritual and human beings are frequently not separated from what non-indigenous people would see as the "natural world". This takes the discussion beyond the more prosaic view of systems of "information" used by outsiders.

Among many indigenous peoples, particularly of the rainforest, specialists establish a technical prowess in production activities and curing illness from their relationship with the spirit worlds. These shamans will often exchange their knowledge of these techniques for goods or services within the community. This knowledge remains as part of the shared heritage of the community, but once it leaves the control of the people, the process of ethnocide takes root.

Shamanic knowledge is recognised as a major repository for information on the uses of plants in the rainforests. About one quarter of all prescription drugs come from the rainforest plants and three quarters of these have been gathered from information provided by indigenous peoples (RAFI, 1989:5). The efficacy of quinine and curare were taught by the Shuar of Equador and Peru. The National Cancer Institute of the USA is currently working on a five year programme to screen plants from all over the world for chemotherapy (Gilbert and Colchester, 1990). Rainforest-derived drugs are now a money spinner of $43 million a year. Hardly anything goes back to the indigenous peoples.

Indigenous peoples have never opposed the use of their plants to save lives. It is the way in which their information is stolen and made into a commodity which angers them. For example, the Uru-eu-wau-wau people in Brazil have a plant known as "tiki uba" which has anti-coagulant properties. The company Monsanto has used this and is attempting to patent the genetic make-up of the plant. The Uru-eu-wau-wau are currently seeking, through the organization Cultural Survival, a way to oppose this commercialisation of their knowledge (Jason Clay, pers. comm.).

A recent publication sets this phenomenon in context:

"In all of this there is a sense of something wrong. Japanese companies are collecting herbs in Asia. American companies are after plants in Latin America. European companies are opening up research centres in Brazil and India. There is money to be made. But none of it will be made by the people who first discovered the value of these traditional medicines (Fowler *et al.*, 1988:166).

Similar problems took place during the Green Revolution during the 1970s. During this period a network of International Agricultural

Research Centres (IARC), which were coordinated by the Consultative Group on International Agricultural Research (CGIAR), organised the collection and conservation of genetic resources. The CGIAR and the Food and Agricultural Organisation (FAO) have established a research institution to conserve and carry out research into genetic resources. The International Bureau for Plant Genetic resources (IBPGR) contains 127 base collections of genetic materials. Eighty-one of these base collections are deposited in the North and a further 29 are controlled by companies from the North. In all, only 17 of these base collections are in the hands of the Third World, even though most of the material comes from there (Shiva, 1989).

At the same time the Union for the Protection of New Varieties of Plants (UPOV) set up a system to compensate breeders and protect rights to varieties. These did not necessarily help the small farmers who could hardly afford to buy the varieties. This situation has been exacerbated recently with the successor of the Green Revolution - the Biorevolution. Here, companies not only have rights of single varieties but over genetic traits within the varieties. These traits can be patented and restricted by companies who monopolise the seed and plant market.

Fears that indigenous peoples are increasingly becoming easy prey for "gene hunters" entering their territories and seeking out plants with specific properties which can be patented and controlled. The current General Agreement on Tariffs and Trade (GATT) negotiations in Uruguay is seeking to promote the patenting of all form of "industrial property". This could further facilitate "gene robbery" (Raghavan, 1990:44).

The proposed Biodiversity Conservation Strategy sees the role of the IBPGR and CGIAR as an important part of their *in situ* (field) and *ex situ* (storage) plans. *Ex situ* storage refers to flora and fauna held in botanic and zoological gardens or in genetic and seed banks. The financial strategy seeks to attract support from private investors. However, we know that these interests are the very same as those trying to control rights over intellectual property and protecting and patenting plant and genetic material (Mooney, 1983 and Kloppenburg, 1988).

There are no built-in mechanisms in the Biodiversity Conservation Strategy to protect the intellectual property rights of indigenous peoples. At the moment the only means they have at their disposal is to utilise the law as it is (such as in the case of the Uru-eu-wau-wau challenging Monsanto's right to patent their traditional medicines). Alternatively, indigenous peoples have to develop their own new form of protecting their intellectual property rights.

A leading exponent of this initiative has been Darrell Posey. Posey argues convincingly that there must be protection of the intellectual property rights of indigenous peoples and they should receive compensation for the contribution they make to the furtherance of knowledge. Posey considers that the World Intellectual Property Organisation could provide this new way (Posey, 1990). In 1984, WIPO proposed "Model Provisions for National laws on the Protection of Expressions of Folklore Against Illicit Exploitation and Other Prejudicial Actions". However, the initiative stagnated because it was too vague and generated little general interest.

A recent report from the Rural Advancement Fund International, 1989, summarises the problem:

"Much more is needed than search and rescue missions from the North motivated by economic interests. Third World countries and indigenous peoples must also benefit from their knowledge and biological treasures. Long-term conservation measures must be put in place. In the process, indigenous people must be treated with respect and given the recognition they deserve (RAFI:1989:7).

Indigenous Peoples and Biodiversity

Dasmann (1988:303) describes the difference between "ecosystem people", such as indigenous peoples whose subsistence comes primarily from the ecosystems where they live, and "biosphere people" who have the whole biosphere at their disposal":

"Local catastrophes that would wipe out people dependent on a single ecosystem may create only minor perturbations among the biosphere people, since they can simply draw more heavily on a different ecosystem...The impact of biosphere people upon ecosystem people has usually been destructive...Biosphere people create national parks. Ecosystem people have always lived in the equivalent of national parks."

The threat to the biodiversity of the world is in the hands of powerful organisations such as the WRI, IUCN and UNEP which have, until now, taken the approach of "biosphere people". The interests of the "ecosystem people" in the area are not a high priority in the agenda. A recent document, published by WRI, IUCN and others, illustrates this point:

"Since local cooperation is essential for the long-term success of conservation efforts, it is usually advisable to undertake a socio-economic survey of the communities affected by projects that involve controlling use of biological resources in order to determine what resources are used, how they are harvested, the degree of awareness about controlling regulations, and possible alternative sources of income.

"Such surveys can also provide the necessary raw material for determining the sorts of incentives required to bring about the desired changes in behavior, as well as the best means of providing incentives and ensuring that they are perceived as fair, equitable, and fairly earned" (McNeely *et al.*, 1990:73).

This quote has the unpleasant flavour of social engineering. Local people are studied in order to get them to change their ways of production to bring about the twin aims of conservation and profit for the engineers.

As if changing indigenous peoples lives were not sufficient, there is further discussion as to the way in which they should change. In the 1990 document of WRI and IUCN, "Gift to the Future", it says that "concerned at the decline of traditional cultures and biological reserves management systems" the strategy proposes to "protect land and models of development that maximise their (traditional cultures') ability to choose the nature and pace of their integration into dominant culture".

For the last five years, the International Labour Organisation has been revising its Convention 107 because its orientation was paternalistic and out of touch with the needs and desires of indigenous peoples. The fundamental problem with the outdated Convention 107 was that it promoted integration of indigenous peoples into the dominant societies.

Integration is not a symbiosis between dominant societies and indigenous peoples, the relationship is asymmetrical. Integration sucks indigenous peoples into the vortex of the national state society - it is nothing less than controlled assimilation.

Although promoters of the Biodiversity Conservation Strategy will say that this is certainly not their intention with regard to indigenous peoples, the effects of their work could be entirely different. This myopia concerning the rights of indigenous and other peoples living in the rainforest leads to two perspectives of the process destroying people and the environment.

From the resource managers' perspective, the world's genetic diversity is under threat and that it is only through *in situ* and *ex situ* conservation that this can come about. In order to finance this, the economic benefits of the rainforest should be used as bait to attract business and financial donors. Indigenous and forest people can be moved from zones of high biodiversity and on "buffer areas" outside of the protection areas they can extract resources and become more integrated into the national economy and society in general.

Alternatively, indigenous peoples' perspective sees the threat to biodiversity as a danger for their continuing livelihood. They consider that recognition of their rights to their territories should be a precondition for conservation as it will enable them to continue their self-sufficient and sustainable lifestyle. In areas where no people live, these can be the protected areas. Any marketing or use of indigenous knowledge should come from their own communities so that they can control their own self-development.

However the problems set out in this paper cannot be simplified into one of "environmentalists" versus "indigenous peoples". On the one hand we have resource managers of the environment who wish to protect and utilise resources. On the other hand we have forest peoples who want to exercise their basic human rights in combination with conservation activities. The environmentalists are thus torn between the alternative of "green capitalism" and "social ecology".

Green capitalism sees the solution to the world's environment problems as arising from an alliance with the market economy. One can thus save the planet and make money at the same time. Social ecology places local people at the centre of solutions to the threat facing the environment. In this way, organisations such as the World Rainforest Movement are seeking alternative strategies using the experiences of local people to create a world where their rights become recognised as inextricably bound with conservation issues (WRM, 1990).

Many environmentalists consider themselves caught between these positions. They agree intellectually with the "social ecological" position, but feel that the only practical solution is to collaborate with those economic forces which have caused the environmental problem in the first place:

"There is no quick fix for the forest. Market forces have driven tropical forests to the edge of extinction. Only market forces can drive their rehabilitation and conservation. The only way to save the forest is to make the trees we want to save more productive, more attractive commercially and more available" (Macklin, 1989:51).

This approach is a recipe for disaster for both cultural and biological diversity because it goes in the wrong direction. The protection of indigenous peoples' rights encourages and facilitates future work and collaboration with environmentalists on conservation matters. On the other hand, if indigenous and forest peoples are destroyed in order to ensure the profitability of the rainforest, they can never be brought back again.

The alternatives presented here are thus not positions whereby one can compromise, taking ideas from here or there. The difference lies in the axioms on which both approaches are based. Indigenous peoples need to exercise their control over the resources of the rainforest and use them according to their basic needs. This has ensured the survival of the rainforest until today.

Top-down approaches are as unacceptable in environmental Conservation as in development. The alternative, namely to work with indigenous peoples, has already been set in motion by organisations such as the World Rainforest Movement and the Coordinadora of the Amazon Basin (COICA) which are stimulating the dialogue between environmentalists and indigenous peoples. However, until those organisations establishing initiatives on the conservation of biological diversity treat indigenous peoples with respect, there will be little hope for either the conservation of biodiversity or the protection of cultural diversity through recognising the fundamental rights and freedoms of indigenous peoples throughout the world.

Note: Since this article was written, a draft of the *Biodiversity Conservation Strategy* has been produced by WRI, IUCN and UNEP with over forty other organisations – none of which is indigenous. In spite of rhetorical shifts in the text recognising

indigenous peoples' rights, the document still betrays a fundamental lack of contact with and understanding of the problems of those people directly facing the threats of biodiversity destruction.

Acknowledgments:

This article is based on the report "Between the Spice of Life and the Melting Pot: Biodiversity Conservation and its Impact on Indigenous Peoples" which was completed in July, 1990 sponsored by the Gaia Foundation and the World Wildlife Fund UK. I thank both these organisations for their support. I would also like to thank the International Work Group for Indigenous Affairs (Copenhagen), Survival International (London) and the World Rainforest Movement (Penang) for providing me with materials. My opinions were stimulated considerably by discussions with Sheila Aikman, David Alexander, Marcus Colchester, Liz Hosken, Anna Lewington, Vandana Shiva, Heffa Schücking and Espen Waehle. Although I am responsible for the views expressed in the text, I would like to thank all these people and organisations for their valuable contributions to this work.

References

Baker, L., 1989 ."Cultural Survival Imports: Marketing the Rainforest". *Cultural Survival Quarterly* 13(3) pp.64-67.

Burger, J., 1987. *Report from the Frontier: The State of the World's Indigenous Peoples.* Zed Books.

Dasmann, R., 1988. "National Parks, Nature Conservation, and Future Primitive" in Bodley (Ed.) *Tribal Peoples and Development Issues*, pp. 309-310. Mayfield, California.

Fowler, C., *et al.*, 1988. The Law of Life: Another Development and the New Biotechnologies. *Development Dialogue.* Uppsala.

Gilbert, D., and M. Colchester, 1990. "Indigenous Peoples and the International Biodiversity Programme: Some Implications". ms. Survival International.

Gray, A., 1990. "The Putumayo Atrocities Revisited", paper presented at Oxford University seminar on the State Boundaries and Indians. ms.

Head, S., and R. Heinzman, (Eds.) 1990. *Lessons of the Rainforest.* Sierra Club Books, San Francisco.

Hecht, S., and Cockburn, 1989. *The Fate of the Forest.* Verso.

Henriksen, G., 1989. "Introduction to Indigenous Self-Development in the Americas". *IWGIA Document 63,* Copenhagen.

IWGIA, 1984. "Tigers versus People", *IWGIA Newsletter No. 40* pp.33-37.

IWGIA, 1989. *Yearbook 1988,* Copenhagen.

Kloppenburg, J., 1988. *First the Seed: The Political Economy of Plant Biotechnology,*1492-2000. Cambridge University Press.

Macklin, D., 1989. "A Nursery for the Rainforest," *New Scientist,* Dec. pp. 47-51.

McNeely, J., Miller, K.; Reid, W., Mittermeier, R., and T. Werner, 1990. *Conserving the World's Biodiversity,* IUCN, Gland, Switzerland: WRI, CI, WWF-US and the World Bank, Washington D.C., USA.

Mooney, P.J., 1983. "The Law of the Seed", *Development Dialogue: Another Development and Plant Genetic Resources.* Uppsala.

Myers, N., 1989. *Deforestation Rates in Tropical Forests and their Climatic Implications.* Friends of the Earth, London.

Pearce, F., 1990. "Brazil, where the ice cream comes from", *New Scientist,* 7 July, pp. 45-58.

Posey, D., 1990. "Intellectual Property Rights and Just Compensation for Indigenous Knowledge: Challenges to Science, Business and International Law." Paper presented to the Association for Applied Anthropology, York.

Raghavan, C., 1990. *Recolonization: GATT, the Uruguay Round and the Third World,* Penang. Third World Network.

Reid, W., and K. Miller, 1989. *Keeping Options Alive: the Scientific Basis for Conserving Biodiversity.* WRI, Washington.

Rural Advancement Fund International, 1989. "*Biotechnology and Medicinal Plants*".

Shiva, V., 1989. *Biodiversity Conservation: The Threat to Ecological Conservation from Commercial Interests.* Third World Network.

Stegeborn, W., 1985. "The Veddas - a people under threat". *IWGIA Newsletter No. 42,* pp. 166-180.

Turnbull, C., 1972. *The Mountain People.* New York: Simon and Schuster.

Wilson and Peter, 1988. *Biodiversity.* Academy Press. Washington, D.C., USA.

Wolf, E., 1982. *Europe and the People without History.* University of California Press.

WRI, IUCN, UNEP, WCU, 1990. *Gift to the Future: Strategy for Local, National and International Action to Save, Study and Sustainably Use the World's Biological Diversity.* Washington.

World Rainforest Movement, 1990. *Rainforest Destruction: Causes, Effects, and False Solutions.* WRM, Penang.

Who Defends Biological Diversity?

Conservation Strategies and the Case of Thailand

Larry Lohmann

How forests, streams and land are treated by social groups depends on what they mean to those groups and how they are organized. Some groups may wind up preserving or enhancing the variety of species, stocks and ecosystems on the land they occupy. Others competing for the same areas will tend to deplete their biological diversity.

Common sense suggests that, other things being equal, scientists, conservationists, international agencies and NGOs can defend biodiversity more effectively by making common cause with the former than by doing so with the latter. Many agencies concerned with conservation, however, have chosen to do precisely the opposite. Slowing biodiversity loss may depend partly on their recognizing this and reversing their stand.

The case of Thailand, where depletion of biodiversity, overlain with various social issues, forms a part of the background to some of the most important national political battles being fought today, provides material to flesh out these claims.

Thailand has seen its forests disappear at a startling rate during the past few decades. Early in this century, up to three-quarters of the country was covered by forests of remarkable variety. Today only scattered patches occupying perhaps 15 per cent of the nation's land area (or less than 80,000 square kilometers) remain to any reasonable degree naturally forested (1).

The biological diversity formerly maintained in the country's agriculture has also declined. Modern livestock breeds have supplanted many village races, and export cropping has come to overshadow traditional rice and mixed agriculture with its variety of local food plant varieties. Green Revolution rice varieties have, in addition, brought with them demands for pesticides and fertilizers that have driven fish out of Thai paddies (2). Soil structure and soil life too have been damaged by chemical inputs. Following a massive expansion of cultivation in recent decades, more than one-quarter of the land is affected by severe or very severe soil erosion (3). The storehouse of traditional medicinal plants, together with the knowledge required to make use of them, has suffered a similar decline.

What has caused these losses? A complex and uneasy alliance between two inter-penetrating, continuously-evolving and powerful groups, both of them urban-based, has been mainly responsible. The first consists of business interests, both national and foreign. The second is a state bureaucracy and army guided by, and allied to, foreign governments and international organizations. Together, these two broad and interlinked groups have progressively usurped control of forests, land, agriculture and water from local villagers.

For the past century and a half, expansion of agricultural land – often combined with timber harvesting – has been a leading cause of forest loss in Thailand. Contrary to myth, this expansion has been driven not simply by population increase but also by the invasion of the market economy and by economic policies handed down from the country's capital. This trend was visible as early as 1855, when Britain, through the Bowring Treaty, pushed the country to increase its exports in exchange for a degree of national autonomy. In the succeeding century, growth in cultivated areas from cleared forests consistently out-stripped population increase, partly as a result of the 25-fold per cent increase in rice exports between 1855 and 1930. Following World Bank and US missions to Thailand in the 1950s, a huge increase in agricultural clearance for non-rice cash crops grown almost exclusively for export or non-local needs accelerated this trend. Between 1950 and 1978, cultivation of cassava, maize, sugarcane and other crops, overwhelmingly for export to Japan, Taiwan, the EEC and other regions, increased 618 per cent, at the expense of perhaps 50,000 sq km of upland forest (4). Repeated monocropping of such plants in areas of low fertility under the patronage of local merchants, moreover, led to loss of nutrients, erosion, weed invasion, lowered yields, the need for expensive inputs, and debt. As a result, many of the migrants who cultivated them became impoverished and were forced to further clear forests elsewhere. Such clearance has meanwhile led to flooding, droughts and siltation downstream. This has displaced other farmers elsewhere, augmenting the vicious circle of impoverishment, migration and environmental deterioration.

When business interests have gained direct access to forests in Thailand, similar devastation has resulted. By 1968, more than 500 logging concessions had been given out covering half the country's land area (5). Regulations calling for replanting, cutting in 30-year cycles, rehabilitation and protection from encroachment were promulgated but nowhere acted upon (though even if they had been, it is doubtful whether they would have done much to conserve biodiversity, since they were not based on careful study of tropical forests). Illegal logging outside concession areas, often backed by sawmill interests, has also been widespread, though more scattered in impact. It was only following catastrophic deforestation-related floods in Southern Thailand in 1988 and the opening up of Burmese timber trade to Thailand that the remaining 300-odd concessions were finally cancelled in 1989.

The corporations now attempting to take control of as much as eight per cent of the country's land area – much of it in former logging concessions – to plant *Eucalyptus camaldulensis* monocultures, are, like the logging companies, interested primarily in the quantity of wood they can extract from a given parcel of land in a short time. They have little stake in maintaining a commercially nonfunctional wealth of species and strains of flora and fauna, nor in preventing farmers resident on their concessions from being forced to flee and clear forest elsewhere. In some cases, eucalyptus firms are even cutting natural forest to provide space for the fast-growing Australian trees (6).

Nor is it in the interest of other types of agribusiness to preserve biological diversity. Tiger-prawn ponds have replaced perhaps half of the country's rich mangrove forests in the space of a decade in order to grow a single species, *Penaeus monodon*, for luxury seafood markets abroad, with crushing effects on local fisheries. The giant Dole pineapple plantation established in Prachuab Khiri Khan in the 1970s with government support and shareholding by top government technocrats is notorious for having wiped out forests, damaged soil fertility and village crops, polluted streams and caused flooding (7). In the same province, Sankyo Ltd.'s plantation of *Croton sublyratus*, a medicinal herb used by the Japanese to treat ulcers, is, ironically, impoverishing the very ecosystem in which the genetic stock used on the plantation evolved (8).

In all such examples, business is merely being true to its own imperatives in the modern world: producing and trading uniform goods on a massive scale while ensuring that as many liabilities as possible, including depletion of biodiversity, are borne by others. The value of a biological organism to a corporation tends to bear little relation to the organism's importance in a local ecosystem or community. Rather, its value lies mainly in its contribution to corporate financial stability or expansion. If something else can take its economic place, a company's natural inclination is to be indifferent to its survival, and even if an organism is economically essential, it is often put at risk in its natural habitat. A firm has little incentive not to deplete the biodiversity in the area in which it is operating, especially, if it can exchange the profits gained from that depletion for different resources elsewhere. The general tendency of biological resources under the stewardship of modern firms is toward extinctions and reduction of ecosystem variety.

In Thailand, as in many other countries, this economics-driven erosion of biological diversity has been greatly helped by the activities of the state and international agencies. Throughout the period of massive agricultural expansion, for example, unwritten government policy was to increase export income and relieve land pressures by encouraging forest clearance rather than through intensification and diversification which would benefit local people in rice-growing areas. Money was put into railroads and roads in preference to agricultural improvements, since these provided more returns to elites and helped extend state control over the countryside (9). A rice export tax has kept rice productivity low at the expense of farmers while adding to state coffers and subsidizing urban wages for the benefit of industrial development. Upland export crops have meanwhile been widely promoted by the government as a means of "diversification" and an opportunity to exploit "comparative advantage".

From 1899 to 1989, the Royal Forest Department (RFD) enclosed land and assisted and regulated logging operations (10). This was indeed the sole purpose for which the RFD was set up; not until the 1960s were national park and wildlife conservation divisions added to it, and they remain small and weak. Today over forty per cent of the nation's land area remains under the control of the RFD, which is now promoting eucalyptus plantations. By taking common forests away from villagers and handing them to concessionaires, the RFD has in effect encouraged villagers to plunder them; as social scientist Shalardchai Ramitanondh puts it, the rational attitude for farmers became: "Why not cut down the trees, when those capitalists from town will definitely come to cut them down anyway?"

Following the first World Bank and US advisory missions in the late 1950s, state-promoted deforestation accelerated drastically as a series of dams and roads was built to foster commerce, suppress insurgency and encourage foreign investment. The dams and roads wiped out some forest directly, including many species-rich lowland riverine areas; six hydroelectric dams alone inundated more than 2000 square kilometers. Infrastructure opened still more to destruction by logging companies, pioneers contracted to open land for large landholding agricultural interests, and other settlers planting the new upland export crops. Total road mileage expanded more than 250 per cent between 1960 and 1980, and paved mileage more than seven times over the same period (14).

Between 1961 and 1985, at least 125,000 square kilometers of forest land were denuded through these mechanisms, about a quarter of the country's land area (15). As crops flowed over the new roads and out of the ports to world markets to help finance industrial development, consumer goods entered the most remote villages. To pay for them, farmers had to sell more produce, plant more cash crops, apply more chemicals and fertilizers, rent more tractors, and borrow more from local moneylenders, often at astronomical interest rates of 5 per cent a month and up. As a result, today, the average Thai farming family's debt exceeds its yearly income. Loss of local control over land, crop choice and other resources has been an inevitable result, increasing insecurity among the poorest and pressures on unclaimed forest land.

Modern large-scale irrigation systems installed under government programs supported by international agencies have meanwhile served to deliver water into the hands of big business, the electricity authority, urban developers, larger landowners and the Army, while progressively taking it out of the hands of the small farmers who have the most interest in, and most experience in, preserving watershed ecosystems. Failing to repay investment, these systems have also caused considerable environmental damage, including flooding (16). Replacement of wooden dams in traditional **muang faai** irrigation systems in the North with "improved" cement structures leads to increased siltation. In addition, the cement structures have often torn apart the complex forest/ stream/rice field/labour relationships which local villagers have maintained for centuries as an ecological guarantee of subsistence. This has sometimes led to abandonment of the system together with the system of forest management which forms a part of it (17), leading to more forest and watershed destruction.

Forest colonization has been exacerbated by land speculation fueled by big business's invasion of rural areas. This has in turn been encouraged by multilateral development banks' increased attention to the economic development of outlying provinces, as well as business-favoring land laws and the growing voice of corporations in recent Bangkok administrations. Increasingly, speculators are persuading small farmers to sell usufructuary rights, then reselling at a profit to resort developers, plantation tycoons or commercial interests who plan to profit from booms created by state-supported industrial estates. In 1989, a total of 3.6 million land transfers was recorded in Thailand – an increase of 56 per cent in a single year (18). Many displaced villagers move on to other forests to establish new farms or just to wait for the next wave of land speculation.

Thailand's state bureaucracy is selectively encouraging precisely those business activities which will accelerate this process – not out of ignorance but following well-considered plans to boost export income and feed the bureaucracy's own needs. Special privileges have been granted to eucalyptus planters, including tax breaks and the opportunity to rent state land for US$1 per acre per year, and many other biodiversity-depleting projects, such as shrimp farms and chemical factories setting up shop in industrial estates, receive public subsidies as well. Land-hungry farmers meanwhile find themselves shut out by land laws, bureaucratic procedures and corruption. Such patterns are simply new expressions of a venerable state ideology according to which forests and land are to be treated primarily as transferrable counters in national economic or political games, rather than as part of the enduring foundation of local village livelihoods.

As in most countries, bureaucracies in Thailand support illegal business activities which deplete biological diversity with hardly less enthusiasm than they encourage lawful depredations. In 1980, for example, then Prime Minister Prem Tinsulanonda ordered a halt in the commercial extraction of salt from subsurface brine in Mahasarakham province following reports that it had rendered a local reservoir saltier than sea water and had contaminated the Sio River. The mining continued, however, with efforts at enforcement consistently stifled by influential people in distant soda ash and glass industries who needed the salt. The river became depleted of fish along its entire 225-kilometer length. As much as 480 square kilometers of farmland was salinized, affecting 300,000 people and inflicting untold damage to local vegetation and soil. Yet letters and petitions to local officials proved fruitless. When villagers finally took to the streets in mid-April 1990 to undertake a last-ditch defense of their ecosystems, they were attacked by 600 armed riot police reportedly sent on the orders of Minister of the Interior Banharn Silpa-archa. Banharn and Industry Minister Pramarn Adireksarn had set up the soda ash and glass companies which were the most prominent beneficiaries of the Mahasarakham salt supply (19). The complicity of high officials in Thailand's Royal Forest Department in the booming illegal trade in rare animals and animal skins is meanwhile well-known to those fighting the traffic around the world.

The direct contribution of the military to the depletion of biological diversity in Thailand has also been significant. During the campaign against the Communist Party of Thailand, the Army actively encouraged villagers and others to destroy forest as a way of denying refuge to insurgents, and still looks with misgivings on remote tree-clad areas. Today, Khor mountain which is one of the last Communist strongholds and covered with rich forest through the 1970s, is treeless and saturated with pesticides from government-promoted cash-crop schemes on its slopes. The need for funds to sustain patronage networks, as well as the need to gain credit with the public in peacetime, moreover, has meant that military commanders take a keen interest in "development" schemes in the countryside, where the Army is a powerful, sometimes overwhelming, force in local politics. Recently, the Army has participated in eucalyptus-planting schemes in the Northeast and has also played leading roles in illegal logging schemes, peasant-displacing land seizures, and timber deals with the military dictatorship in Burma. The latter alone have denuded hundreds of square kilometers of undisturbed primary forest in the last two years and promise to ruin much more.

In addition, military units are also active in efforts to suppress tribals, whose societies are already under severe ecological pressure due to the migration of millions of marginalized lowland Thais into the hills of the heavily-forested North. In 1988, for example, the para-military Border Patrol Police, a unit which had served to absorb potentially disaffected rural youth into the power base of its patron, former Army Commander-in-Chief Chaowalit Yongchaiyudh, burned villages and fields of the Akha tribe and drove the residents over the border into Burma. The military have also participated in the forced resettlement of Hmong and other tribal peoples from watershed and other protected areas onto infertile, poorly-watered or otherwise inappropriate land which, as experience has shown, cannot support them.

Due to these factors and others, it is increasingly difficult for the Akha, Hmong and other tribes to live in the customary ways which provided substantial protection for biological diversity. Due largely to pressure from lowland Thais, security worries, and lack of territorial and citizenship rights, there remain few Akha villages in Thailand where traditional forest belts around villages (maintained for protection against marauders and as a source of game and other essentials) can be preserved, and knowledge of medicinal herbs is disappearing nearly as fast as the forest which supplied them. Threats to tribal societies such as those found in Thailand are an especially powerful threat to biodiversity, given the close links between cultural diversity and biodiversity and the fact that 90-95 per cent of the world's cultural diversity is to be found in tribal societies.

II

If corporations and the state, backed by international agencies, are the principal destroyers of biological diversity in Thailand, who is most effectively defending it? Despite repeated shows of concern by the government and several official programs to mitigate environmental damage, events of the last five years suggest that, by and large, biodiversity's most powerful national defenders are ordinary farmers – untutored in academic biology and sometimes innocent of the word "enviromentalism" – with occasional support from the media, non-governmental organizations and maverick government scientists.

To say that grassroots movements are by their nature more effective than government or corporations in defending biological diversity, of course, is not to say that all or most rural villagers – who in Thailand constitute three-quarters of the population – are environmentalist, nor even that village activists do not also contribute to biodiversity depletion. The economic pressures of modernization have pushed many villagers into participating in, as well as occasionally resisting, the destruction of biological diversity, just as they have pushed most Westerners of all shades of environmentalist and anti-environmentalist opinion into doing so. It was villagers, after all, who constituted the bulk of the work force of the teak-extracting British Borneo Company a century ago. It is villagers who are hired by middlemen to poach the rare animals which find their way to urban or foreign markets. It is villagers who are contracted by local firms to degrade forests to provide excuses for companies to ask the state for the right to rent them. And it is, of course, villagers who, motivated by the need to regain their own means of subsistence, by opportunities for new income from export crops, and by population growth, have pushed forest frontiers back from original settlements in river valleys.

Where solid positive steps have been taken, however, they have generally been either taken by villagers directly or forced on government or corporations by pressure from the village level. The reasons for this are not difficult to find. The same economic pressures which suck villagers into cycles of environmental degradation eventually also threaten directly the environmental sources of rural livelihoods. At various times and places, villagers may be forced into migrating into new forests along new roads, hired by sawmills to supply illegal wood, or induced to attempt to boost their status through consumption or participation in government development projects. But if business or the state go too far in seizing or undermining environmental bases of subsistence to which villagers feel entitled, they will act. This point is being reached along several fronts today as forests, remote watersheds, and recently-cleared land which used to be of peripheral interest to elites are drawn into their economic orbit. Areas which formerly supported self-reliant village agriculture or acted as refuges for the disaffected or marginalized are increasingly becoming battlegrounds where social classes fight over who has the right to use and benefit from the country's dwindling areas of high biological diversity.

Villagers' battles against hydroelectric dams are a case in point. One by one, the national electricity authority's takeover of forests and streams to power the industrial and commercial sectors with electricity began years ago to pinch provincial populations to the point of protest. Local opposition to resettlement schemes connected with dams was visible as early as the 1960s. Later, as more and more villagers started to suffer from the destruction of local natural support-systems which the dams entailed, dissent grew. In 1988, local dam opponents, backed by students, environmentalists, and the media, succeeded in forcing the government to shelve the Nam Choan project in Western Thailand. Prominent among objections to the dam were claims that previous dams, through deforestation or other mechanisms, had adversely affected local climatic conditions, soil fertility, water and fishery resources. Local people were also concerned that one of the largest remaining tracts of undisturbed forest in mainland Southeast Asia would be threatened by dam construction and subsequent colonization. Biologists concerned about the conservation value of the forest that would be lost learned quickly that, reinforced by grassroots pressure, their voices were carrying more weight with policy-makers than ever before. Alliances were thus set up between conservation-oriented scientists, local groups, students, and the media that are still growing and serving the environmental movement (20).

The years since have seen an explosion of popular resistance to other dams planned across the country. Villagers marching in the streets have stalled the progress of projects aimed at harnessing the Mool and Yom rivers in the Northeast and North. And in July 1990, heavy local opposition forced the government to retract a Cabinet decision giving the go-ahead for the Kaeng Krung dam in a primary forest in Thailand's South. In June, villagers from existing and proposed dam sites across the country had met in Bangkok to demand a nationwide moratorium on all further dam construction. Conservationists, development workers, professors, and other city-based intellectuals have increasingly lent their voices and expertise to this movement, with their concerns reaching an increasingly wide audience as a result.

All of the dams being protested would cause significant and wholly unnecessary destruction of biodiversity, both through their direct impact on forests and streams and through their displacement of villagers into more marginal areas. Despite the threats posed by these projects, however, opposition arising from within the government, business or universities alone has never given dam builders as much pause as the current movements consisting of large numbers of ordinary farmers.

The most powerful curbs on logging, too, have been those arising from resistance by farmers who cannot afford to see sources of water and other essentials disrupted. Many Northern villagers in particular have long opposed commercial timber harvesting in watersheds in the hills overlooking their rice terraces, on the ground that it dries up streams and silts waterways so extensively that it is impossible to maintain traditional irrigation systems. Such logging also tends to prevent villagers' access to important wild sources of game, vegetables, fodder, fruit, bamboo, mushrooms, insects, firewood, house construction materials and herbal medicine. In addition, it disturbs local graveyards and the residences of spirits of place.

For these reasons, some Northern villagers, having exhausted their patience with fruitless appeals to the authorities, were blocking logging roads at least 15-20 years ago. As deforestation accelerated during the 1970s and 1980s, villagers became more desperate. Farmers tried to stop forestry officials from stamping trees for felling, held rallies, ordained trees in Buddhist monks' robes to deter loggers from cutting them down, and occupied local forests to stop timber companies from removing logs from them. They also enlisted the support of students, scientists, environmentalists, lawyers, development professionals and the media and began linking up with and learning from farmers in other parts of the country fighting similar battles. By 1988 they were in a powerful enough position nationwide to contribute significantly to the successful campaign to pressure the government into banning logging.

One village scored a further precedent-setting victory a year later in its struggle to preserve a local forest against encroachment by an influential Member of Parliament and businessman. As Witoon Permpongsacharoen of the Project for Ecological Recovery tells the story,

> "(t)he MP had rented part of the forest from the government, its nominal owner, in order to construct a tourist resort. The area in question, however, was regarded as a community forest by the local villagers, who pointed

> out that its status as a traditional source of water and food was being threatened. As a result, they marched, petitioned, forced the MP's workers off the land, and attempted legal action. One of their number was assassinated in the course of the struggle, but in the face of unyielding protests, the government finally suspended the rental agreement and admitted the right of the villagers to look after the forest themselves. This constituted the first official recognition of rights of local communities to forests on government land" (21).

In a somewhat different type of battle, it was again local people (including local government officials) who spearheaded opposition to a move by a state timber enterprise to reopen inactive logging concessions in the Huai Kha Khaeng Wildlife Sanctuary in the late 1980s. Motivated by conservationists' concern about the fate of the last primary forest in their province, as well as by worries about climate change, the local residents were among the many groups across the country which joined forces to campaign for the 1989 logging ban, which bestowed a reprieve on the sanctuary. Again, conservation-oriented scientists found that by assisting this already-existing movement, they could more effectively make their own concerns felt in high circles.

Some of the most impressive village-based movements on the Thai scene today are those which are challenging proposals to turn much of the country into eucalyptus monocrop. Biologically-impoverished eucalyptus groves cannot provide the range of foods, medicines, fodders and building materials which many villagers on the edges of the market economy need to fill out their subsistence and which can be supplied by patches of secondary forest. Resistance in many areas is accordingly centered on villagers' need to retake control of local areas to restore more varied ecosystems. Ban Sio in semi-arid Sri Sa Ket province provides one example. More than 40 years ago, local villagers there experienced drought and loss of soil fertility when the surrounding forest was cut and planted with jute and cassava. They then decided to try to rebuild the local ecosystem by allowing the forest to regrow. It did, and is now a source of fodder, herbal medicine, firewood, protein, mushrooms and vegetables – particularly valuable for the poorer villagers. In 1971, however, the area was declared a National Forest Reserve. This gave the forestry bureaucracy the power to lease it to a private company to plant eucalyptus, which it did in 1985, giving as an excuse the claim that the forest was "degraded". When bulldozers moved into the forest, local villagers responded en masse by burning down the concessionaire's office and chopping down eucalyptus saplings. Later, they were forced to put a heavy guard on their village to protect against intimidation, assassination and reprisals, but no more clearing of forest or planting of eucalyptus has taken place since then, and Ban Sio villagers are continuing to use and protect their forest as it slowly regrows. But the concession contract is still legally valid, and the farmers remain watchful.

Elsewhere, villagers resisting eucalyptus are attempting to pre-empt or replace the plantations with trees of their own. Significantly, these are typically not monocultures but rather mixed stands of native, fruit or rubber trees. On Environment Day, 1990, for example, people from eight provinces fighting eucalyptus came together to plant trees in a single location in Yasothorn. The action was a variation on an old cultural ritual in which villagers band together to help construct or refurbish Buddhist temples.

Some farmers, with support from progressive community hospitals and non-governmental organizations, are also attempting to restore the traditional place of native medicinal herb gardens in village life. This particular move toward rebuilding local biological diversity came as a response to the feeling that dependence on the wider market economy for medicine had led to both financial insecurity and health risks. A decade ago Ban Sok Khumpoon in Yasothorn province, which, together with many such villages in the Northeast, had been swiftly integrated into the cash economy in the 1960s and 1970s, was spending over US$500 a year on commercial drugs. These included pain-relievers whose use led to the deaths of several residents from stomach ulcers. Now, however, following years of community efforts, medicinal herbs are growing again in quantity around the village, the drug bill has been cut by three-quarters, the village is greener, and local peoples' health is better (22).

Such attempts to restore village self-reliance through rebuilding local biological diversity go far beyond the realm of medicine. In many places around the country, villagers whose lives have been made insecure by overdependence on the market economy have become increasingly wary of single-crop cultivation and its requirements for costly and dangerous chemical inputs. A few farmers are experimenting with organic mixed farming and agroforestry in search of ways of securing their families' subsistence without relying excessively on the market economy. They raise dozens of types of crops, trees and livestock together in carefully-managed systems, and are attempting to return fish to the rice paddies. They supply most of their own needs without money yet are capable of producing small cash surpluses at any season to meet emergency needs.

Such efforts are important for biodiversity preservation not only because they help bring back variety and fertility to the village ecosystem. More significantly, they help indirectly, by reducing the threats of indebtedness and landlessness and enriching and reintegrating community life. The need to move to the cities or to colonize unoccupied forests elsewhere thus diminishes, along with the pull toward adoption of unsustainable urban lifestyles.

One example famous within Thailand is that of Wiboon Khemchalerm of Ban Huaihin in Chachoengsao province. Years ago, as the owner of a large cassava farm, Wiboon found himself caught in a hopeless downward spiral of debt. As crop prices fell, inputs increased in cost. It became necessary to sell increasing amounts of cassava just to stay afloat year to year. That meant pushing the land to produce more and more, using more and more chemical fertilizers and pesticides. The result was ecological as well as economic and spiritual degradation.

In the end Wiboon decided to sell most of his land to pay off his debts and start afresh. On the tiny one-and-a-half hectare plot that remained, he began planting a variety of crops, trees, and medicinal herbs, eschewing chemical help and distancing himself from the consumer economy. As Wiboon puts it, "I decided to farm for security in life instead of profit."

He became an expert and adviser on herbal medicines. Within five years, his hectare and a half contained more than 500 species of trees and other plants, and small animal life too is on the rebound. For Wiboon, this diversity is connected with security and peace of mind, which in turn is connected with a return to Buddhist values and comparative independence from the market. "Life has become much simpler" , Wiboon says. "I feel contented and secure. I regret that I didn't think of this before I lost my land." It is significant that the restoration of biological diversity associated with Thai programs such as Wiboon's is typically not connected with attempts to find an industrial market for the products of biologically diverse areas. Rather, it is tied to conscious and creative attempts to **disengage** from the wider market economy, and to put it in a subordinate place in social and cultural life (23).

Other Thai farmers who have not gone as far as Wiboon did in embracing modern agriculture have not had to start from scratch in their efforts to rebuild biological diversity. Hundreds of villages in the North, Northeast and South of the country already have community forests which they have carefully maintained over the years. The regrown secondary forest at Ban Sio and the watershed forests of villages in the North are only two examples. And as villagers in many provinces come to the conclusion that industrialization and modernization are not likely to remove the need to maintain older patterns of gathering and ecosystem maintenance as security for poorer farmers, they are eager to extend such locally-controlled woodlands (24).

Each village has to evolve its own specific model for community control, based on local climate, water regimes, species, and social conditions. Farmers in Ban Toong Yao in Lamphun province, for example, have over 60 years developed a set of written community laws which dictate how the local forest is to be used. Trees can only be cut for genuine necessities, such as to build houses for newlyweds. Those who chop down trees for sale on the market or for other purposes face penalties handed down by the village government. These penalties cannot be treated as mere "costs" by budding timber enterpreneurs, moreover, as they might be within the modern market economy, since they also involve a considerable social stigma. The group of villagers who govern the local traditional irrigation system also inspect the forest, keeping tabs on who is using it, for what, and when, and preventing outsiders from exploiting it. But all villagers are responsible for doing whatever is necessary to ensure that the forest is protected as a source of water, food, medicine and wood. In carrying out this task, Toong Yao villagers do not desire the assistance of the government. "All the government has to do is recognize our rights to it", says one leader. "We want to take care of it ourselves, and we can do that better than anyone else" (25).

Critics point out two important limitations in such village conservation efforts. The first is that their impact has so far been small in comparison with underlying trends involving logging, extension of land under commercial agriculture, economic expansion, colonization and environmental degradation. While this is undeniable, it is necessary to add that village initiatives, by their very nature, show greater potential than other programs aimed at rehabilitating degraded forest land or checking further colonization. Official programs invariably fail due to the lack of fit between informal local tenure systems to which villagers remain committed and the land-use patterns the government attempts to impose, as well as other factors. They have "often resulted in increased immigration and forest deterioration" (26), while "legislation controls over land ownership have been a significant factor in the accelerated depletion of forest resources" (27). Internationally-backed programs, while occasionally more successful, are difficult to replicate due to high cost (28). Commercial "reforestation" efforts are meanwhile widely acknowledged to have been counterproductive so far in terms of conservation.

A second limitation of village conservation efforts is that they typically do not embrace all of the concerns of wildlife biologists. The sort of biological diversity Thai farmers want for their own local areas tends to be of the moderate level which can support their ivelihoods (although many with direct experience of climatic changes or degradation of water and soil regimes are also strongly in favor of maintaining intact forests at a considerable distance from their fields). Farmers tend to have less interest in the conservation of specific species not in day-to-day use or in the establishment of strictly protected areas from which they are banned. In addition, they may be indifferent to the issue of establishing protected areas of the large size required for comprehensive biodiversity protection. Their ultimate objectives may thus differ from those of conservation managers intent on preserving areas of extremely high diversity in an untouched state.

In the short term, however, committed conservation-oriented scientists are more likely to be able to form fruitful strategic alliances with village-based movements than with old university classmates who are now working in government electricity authorities, or the World Bank. This is because, as has been suggested above, many grassroot groups tend to have an immediate stake in defending the practices that conserve some level of biodiversity and self-reliance, while staff of government ministries, international agencies or corporations tend to have a vested interest in encouraging the policies that lead to vicious circles of environmental degradation and impoverishment. Village movements which are concerned with issues of land rights, for example, inevitably reach deeper into the roots of deforestation including deforestation in protected areas – than official approaches which confine themselves merely to (say) National Park policy, sponsorship of research, or suppression of encroachment.

Conservationists who want to attempt to reform the approaches of bureaucracies and the market economy to biological diversity, moreover, have found that that is all the more reason for them to devote more effort to understanding and attempting to back village movements. The Thai experience indicates that reasoned conservationist argument sways officials and businessmen only when combined with demonstrations and uprisings at the grassroots or the threat thereof. Mutually-beneficial alliances between groups who understand and are respectful of the differences between each others' interests and motivations have grown more and more important in the Thai environmental movement in recent years.

This is true even with regard to protected area policy. Only after public pressure resulted in a logging ban were more protected areas gazetted. Only by allying themselves with local groups were conservation-oriented scientists effectively able to bring their concerns about Western Thai wildlife sanctuaries threatened by dams and logging before the government in the late 1980s. Today, due to the broad-based movement which resulted, these sanctuaries have won reprieves, enabling scientists and conservation organizations to undertake work on long-term plans for conserving them.

III

Experience suggests, in sum, that in Thailand it is rural village groups and movements opposed to the schemes of governments, corporations and international agencies who are the most powerful and committed defenders of biological diversity. A good test, then, of the probable effectiveness of current global conservation proposals would be to ask whether they would be able to take account of this fact.

A look at the literature on such proposals indicates, unfortunately, that they would not. All assume **a priori** that conservation must take place within a neutral, all-encompassing social matrix consisting of the expanding global market economy, existing relations of state-village authority, and official national development programs; that, in short, "conservation can be regarded as a form of economic development" provided for "through appropriate policies from the central government" (29). The views such proposals take toward villagers' resistance movements and systems of local control are reminiscent of those of the grandmother John Holt describes, who, whenever she heard an unfamiliar notion, "responded in one of two ways: (1) it is crazy, or (2) I've always known it" (30). Most planners tend either to dismiss grassroots conservation activism as an illegitimate topic for consideration or to misinterpret it as part of the economic development they have "always known".

The World Bank's incipient new forestry policy is a good example of the first response. There we are told that conservation is being thwarted because biodiversity, since it typically does not appear in "national income accounts", is "undervalued" (31). Think tanks in the US and Europe, similarly, are producing books and papers on how to persuade governments and local people to "recognize" the value of biodiversity by looking into its economics (32). The fact that biodiversity is in fact already highly valued by communities such as Toong Yao who depend on it for subsistence is ignored; non-economic systems of valuation and the conservation movements to which they give rise are implicitly regarded as simply beyond the pale.

The notion that the political leadership provided by grassroots groups in countries like Thailand might be central to conservation movements is also treated as if it were too exotic even to mention. There are probably several reasons for this. One is that many conservation planners interested in Southern countries tend naturally to try to become advisers to elites in governments and international agencies, out of a natural affinity with these elites' culture and a belief in their potential positive influence. According to *Conserving the World's Biological Diversity*, a volume published in 1990 by the World Bank and four environmental organizations, for example, the interesting actors are not villagers opposing encroachments of the market economy and development programmes, but rather their adversaries: industry, commerce, the state, the military, concessionaires, international conservation organizations, the World Bank, USAID and the Bank-housed Consultative Group on International Agricultural Research (33). In the experience of many Western planners, moreover, conservation is associated more with protected areas funded by the biodiversity-destructive urban economy than with maintenance of symbiotic relationships between rural livelihoods and local forests and streams and the necessity of preventing dispossession. They thus tend to ally themselves with potential funders of protected areas while neglecting, say, land tenure issues or the practices of farmers such as Wiboon Khemchalerm.

Such alliances require that planners avert their gaze from the biodiversity-destroying aspects of elite rule. In 193 double-column pages, *Conserving the World's Biological Diversity* neither acknowledges, nor recommends, countering the threats to biodiversity posed by dams, roads, logging, tree plantations, trade policies, and programs that lead to forest colonization. Instead, it advocates what amounts to a Sisyphean exercise of implementing scattered, top-down projects in their wake – corporate-funded programmes, schemes to siphon a bit of funding out of large diversity-depleting development projects for "mitigation" or watershed protection, foreign ecotourism, debt swaps, capital investment in biological resources, set-asides, and projects which give "total management control" to the "private or NGO sector" (34). Nor does the book criticize military regimes which are devastating the environment. On the contrary, the Burmese Army, together with several others, is described as "having a positive influence on conservation of biological resources" – one which can be enhanced by training programs for "impressionable post-adolescent male" recruits from a rural background who "have a particular affinity for nature and the outdoors" (35). It would obviously be inconsistent with this overall approach for planners even to recognize the existence of village conservation movements opposing development projects or military exploitation, much less to acknowledge their effectiveness in pressuring governments and corporations.

Where village movements do become visible in planning documents, they tend to be reduced to the role of obedient lieutenants in externally-directed programs. In *Conserving the World's Biological Diversity*, for example, local people appear less as creative agents in conservation efforts than as relatively passive adjuncts whose "cooperation" and "collaboration" should be secured and who must be "provided with tools" for conservation by outsiders (36). In a more subtle attempt to dismiss the importance of grassroots groups, many planners insist that they are inherently weak forces which must be instructed and "empowered" by outsiders. This type of political understanding would obviously be incapable of coming to terms with recent events in Thailand, where if anything it is villagers who have "empowered" NGOs and other conservationists in the nation's environmental movements.

Another way in which planners attempt to domesticate village conservation movements is to treat them as if their practices and values were assimilable to those of the market economy. This enables them both obstensibly to advocate "maximum possible cultural diversity" (37), and, at the same time, to endorse expanded roles for the market and technocratic management in peasant and tribal societies. Thus, while local knowledge is praised, it is treated not as something villagers use to defend biological diversity in their own social context but rather as raw material to be translated "into forms useful to development planners and managers of biological resources" (38).

Correspondingly, the knowledge held by village women is to be "mobilized" for universal use through being translated into and "recognized" by the neutral language of Western science. Local practices are to be given an economic meaning as "appropriate prices" are assigned to biological diversity and "cost-benefit analyses ... carried out as a basis for designing incentive systems" (39). Authorities can then "bring the benefits of conservation" (40) to local people by choosing those aspects of traditional knowledge that "can contribute to conservation and development" (41). Similarly, it is assumed that any incentives local people might have for conserving biodiversity must be based on the market economy and government intervention. According to the book, it is not local systems besieged by modernization but rather "prosperity" to be achieved through the market economy which will enable communities to "protect the resources" (42).

Such a redescription would distort Thailand's most significant village movements beyond recognition. The market economy there is not a neutral medium in which a variety of village conservation activities can be lovingly cultured but rather a corrosive acid bath which dissolves most of the diversity-protecting practices it comes into contact with. Villagers' incentives to defend biodiversity in Thailand are not centered on anticipated market or development benefits, but on subsistence guarantees (as in the case of struggles over land, plantations and traditional irrigation systems), long-term safety and security (as in anti-dam campaigns), or the need to preserve independence (as in some community forest and ecological agriculture movements).

Even if the more flagrant abuses of developers and bureaucrats could be checked, there is no reason to believe that industrial society could either assimilate or preserve within itself the conservation practices of a vernacular society such as Ban Sio or Ban Toong Yao. Those practices derive their significance and viability from being part of a web of other practices specific to that society. To transfer those practices or their accompanying theories into a society dominated by market economics would be to attempt to put them into relationship with a new set of practices. This would result in either the practices and theories being regarded as quaint or lacking in point or in their being transformed into practices indistinguishable from those well-established in industrial society. As any peasant or tribal people knows, either result entails violence to local ways, including local ways of protecting biodiversity.

Current global conservation plans, then, seem structurally incapable of making room for the movements which in Thailand offer the best hope for the maintenance of biodiversity. They will remain so at least until they discard the assumption that the market, national development plans and top-down management constitute a neutral background against which all efforts to preserve biodiversity must take place, and recognize that neither industrial society nor any other society is capable of providing such an all-encompassing matrix.

Because we are all a little like the grandmother Holt describes, however, that will not be practically possible until the full diversity of village societies and movements is actually represented in any discussions on comprehensive conservation programs. If there is to be an effective approach to conserving the world's biological diversity it must be one which is developed by all the world's peoples – by hilltribe villagers as well as bureaucrats, by small farmers as well as professional foresters. Not presuming to speak for others is also one of the principles of the conservation of diversity.

Thanks to Witoon Permpongsacharoen and Heffa Schücking for helpful pointers and comments.

NOTES

1. Philip Hirsch, "Deforestation and Development in Thailand," *Singapore Journal of Tropical Geography:* 8 (2) (1987), pp. 129-138; Norman Myers, *Deforestation Rates in Tropical Forests and Their Climatic Implications* (London, 1989), pp. 40-42.
2. Cary Fowler and Pat Mooney, *Shattering: Food, Politics and the Loss of Genetic Diversity* (Tucson, 1990), p. 217.
3. Thai Development Research Institute, *Thailand Natural Resources Profile* (Bangkok, 1987), p. 38.

4. Harald Uhlig, *Spontaneous and Planned Settlement in Southeast Asia* (Hamburg, 1984), pp. 123-124.

5. Komon Pragtong and David E. Thomas, "Evolving Management Systems in Thailand" in Mark Poffenberger, ed., *Keepers of the Forest* (West Hartford, Connecticut, 1990), p. 172.

6. "Suan Kitti implicated in new logging scandal", *Bangkok Post,* 31 January 1990.

7. Justice and Peace Commission, "The Dole Pineapple Plantation", mimeo, Bangkok, 1978.

8. "Plao noi ... samun phrai thai phuea prayote khong khray?", *Warasarn Samun Phrai* 1990, pp. 31-40. Thanks to Rosana Tositrakul for calling my attention to this article.

9. Philip Hirsch, *Development Dilemmas in Rural Thailand* (Oxford, 1990), p. 46.

10. Kamon and Thomas, op. cit., pp. 168-169.

11. Mark Poffenberger, "The Evolution of Forest Management Systems in Southeast Asia" in Poffenberger, ed., op. cit., p. 8.

12. Witoon Permpongsacharoen, "Widerstand gegen Waldzerstorung-Lektionen aus den Dorfen Thailands" in *Arbeitgemeinschaft Regenwald und Artenschutz,* Naturerbe Regenwald (Giessen, 1990).

13. Nart Tuntawiroon and Poonsab Samootsakorn, "Thailand's Dam Building Programme: Past, Present and Future", in Edward Goldsmith and Nicholas Hildyard, eds., *The Social and Environmental Effects of Large Dams,* vol. II, pp. 291-303.

14. Hirsch, op. cit., p. 50.

15. Philip Hirsch, "Forests, forest reserve, and forest land in Thailand", in *The Geographical Journal* 156 (2) (1990), pp. 166-174.

16. Dr Prakorb Wirotnakoot, *"Khuen kap karn cholprathaan"*, presentation at a seminar on dams, effects and energy solutions, Chulalongkorn University, 10 June 1990.

17. Chatchawan Thongdeelert and Larry Lohmann, "The Muang Faai Irrigation System of Northern Thailand", *The Ecologist,* March/April 1991.

18. "Leh' nak long thoon kwaan sue thii pan rakhaa chaaw naa lome lalaai", Krungthep Thurakit Sapdaa, March 1990.

19. Ecumenical Coalition on Third World Tourism, "Death of the Siew River: Chronology of a Protest" (Bangkok, 1990); "Salt in the Wound", Far *Eastern Economic Review,* 7 June 1990, pp. 28-29.

20. Philip Hirsch and Larry Lohmann, "The Contemporary Politics of Environment in Thailand," *Asian Survey* 29 (4) (1989), pp. 439-451.

21. Witoon, op. cit.

22. Sanitsuda Ekachai, "Pa pah tree ceremony gives hope to villagers", *Bangkok Post,* 28 March 1989.

23. Wiboon Khemchalerm, *Saai phan chiiwit* (Bangkok, 1986).

24. NGO Coordinating Committee on Rural Development, "Panhaa paa maai thii tham kin lae thaang awk" (Bangkok, 1989); Sanitsuda Ekachai, *Behind the Smile* (Bangkok, 1990).

25. Montree Chanthawong, *"Paa choom chon"* (Bangkok, 1989).

26. Kamon and Thomas, op. cit., p. 180; James A. Hafner and Yaowalak Apichatwallop, "Migrant Farmers and the Shrinking Forests of Northeast Thailand" in Poffenberger, ed., op. cit., p. 216.

27. James A. Hafner, "Forces and Policy Issues Affecting Forest Use in Northeast Thailand 1900-1985" in Poffenberger, ed., op. cit., p. 85.

28. Hirsch, op. cit.

29. Jeffrey McNeely, "What Value for Wildlife?", I*UCN Bulletin,* December 1990, p. 15; World Bank, World Resources Institute, International Union for

Conservation of Nature and Natural Resources, Conservation International and World Wildlife Fund, *Conserving the World's Biological Diversity* (Gland, 1990), p. 125.

30. John Holt, *Escape from Childhood* (Harmondsworth, 1975), p. 17.

31. World Bank, *Forestry Policy Brief* (Washington, 1990), Annex 4, p. 1; *Forest Policy: An Approach Paper* (Washington, 1990), pp. 13-14.

32. E.g., Jeffrey A. McNeely, *Economics and Biological Diversity* (Gland, 1988).

33. World Bank *et al., Conserving the World's Biodiversity,* pp. 115, 118-126, 131.

34. Ibid, p. 126

35. Ibid, p. 131.

36. Ibid, pp. 73, 105, 109.

37. World Bank *et al.*, op cit., p. 15.

38. Ibid, p. 114.

39. Ibid, p. 113.

40. Ibid, p. 52.

41. Ibid, p. 114.

42. Ibid, p. 35; McNeely, op. cit., supra 29, p. 15.

Genes for Sustainable Development

Overcoming the obstacles to a global agreement on conservation and sustainable use of biodiversity

David Cooper

Introduction

Negotiations towards a global biodiversity convention began at the end of February, 1991. UNEP, which is facilitating the negotiations hope that a framework convention will be ready for signing at the UN Conference on Environment & Development in June 1992 in Brazil. However, many of the fundamental principles have yet to be agreed upon.

This article looks at the likely shape of the convention, and discusses in detail some of the main sticking points. The convention is likely to be successful in addressing the problem of biodiversity loss only if it:

- guarantees access to genetic resources;
- provides remuneration to developing countries, including concrete support to farmers and other people's organizations directly involved in biodiversity conservation;
- contains an agreement on access to the new technologies.

Everyone agrees that the world's biological resources are under threat. Destruction of tropical forests and other ecosystems, the changing global climate, and increasing replacement of indigenous agricultural systems by "modern" uniform ones, are just a few examples of the forces behind it. In an effort to turn the tide, the UN Environment Programme is embarking upon the challenging task to draw up a global convention on biological diversity. The aim is to have the framework ready for signing by the UN Conference on Environment and Development in 1992 in Brazil. However, there are many different ideas on what such a convention should entail and even some of the basic concepts remain controversial.

GRAIN, as an independent international NGO, has been working closely on biodiversity issues for many years, especially in relation to the conservation and use of plant genetic resources and the role of small farmers in sustainable development. It is from that focus that we look at the proposed convention and possible ways forward to overcome the obstacles to reach an agreement which ensures that future generations will have access to the full range of global biological resources for their survival and development.

Conservation and Use of Biodiversity: Contradiction or Synergy?

Substantial new global measures are needed to promote conservation and sustainable utilization of plant genetic resources. Many of the problems to overcome stem from the questions: Conservation for what purpose? Utilization for whose benefit?

Conservation and development have for long been seen as competing and contradictory. Conservationists emphasised the need to conserve natural resources in protected areas – national parks and the like – in which human economic activity would be excluded. Others, maintaining that human and economic development were paramount, claimed that Third World countries in particular could not afford the luxury of nature conservation when so many human needs remained to be met. While many types of development are inimical to the purposes of conservation and vice versa, the last decade has witnessed a major change in attitudes as both sides have begun to realize that conservation is required for sustainable development, and that sustainable development is required for conservation. However, while there is a growing realization of the synergy between conservation and development, differences remain, and there is still a long way to go before they are fully integrated into economic decision-making.

Differences between conservation and development are often confounded with differences in North-South perceptions. The developing countries are understandably concerned that their socio-economic development is not held up by people in the North who are more concerned with nature conservation. These differences were clearly visible during the negotiations to establish the UN Conference and Environment and Development and can be seen too in the UNEP negotiations for a convention on biodiversity.

In the context of biodiversity conservation, the industrial countries have, in the main, put most emphasis on measures for *in situ* conservation, demanding the setting-aside of protected areas. The economic reasoning behind this is they want to ensure that there is a continued supply of components of biodiversity — actual and potential raw materials — for plant breeding, for new medicines, and for the wider biotechnology industry. Many developing country governments, on the other hand, question why they should forgo development of their own territories simply to guarantee conservation for the benefit of the North which some Third World groups call "commercialized conservation". These groups argue that Third World governments should revoke their sovereign rights over their biological resources, restrict access to commercial interests in the North, and develop the resources for their own national benefits.

The problem with over-emphasis on conservation of biodiversity for its components is that those who have benefitted most from their utilization, are not the owners or "curators" of this biodiversity, but those who possess the technology and have access to the markets required to develop and to profit from them. This situation could be redressed by providing protection and remuneration to the providers of the genetic resources, and by facilitating the transfer of relevant technologies to enable the source countries to profit directly from their biological diversity. These two issues are addressed later in this briefing.

But linked to conservation of biodiversity is its use in production systems. Many argue that only agricultural, forestry and other production systems which require and sustain biodiversity can provide direct incentives for maintenance, and thereby guarantee its conservation. Unless diversity is used in production, its conservation will always be under threat from other pressures on land use. The key to maintaining such systems is the experience and knowledge of farmers and other local peoples who have been the curators of such systems over generations. Maintenance and development of these systems is required not only to conserve a pool of valuable genetic material, but it is also vital for their food security and as a basis for sustainable development. This is the third issue which this briefing examines in detail.

The lead up to a new Convention

After several sessions of an expert working group, an initial negotiating document has been formulated and will be the starting point for negotiations which began in February 1991 in Nairobi. Many issues are still to be resolved, but the overall shape of the proposed convention can be deduced and is presented, in a summarised form, in the following pages.

At this point, the negotiators agree that a convention with measures for concrete action is urgently required incorporating mechanisms to promote both conservation and sustainable utilization of biodiversity. It is also agreed that a firm commitment on financing is essential and that those who enjoy most benefits must contribute most. Most consensus exists on conservation measures required at the national level.

As well, there is formal agreement that an innovative mechanism for access to resources and new technologies should be included in the convention, and that these should be based on mutual agreement and reflect sovereignty over natural resources. This point, however, hides a profound disagreement. The rules governing access to genetic resources and technology are amongst the fundamental issues still to be resolved.

Put simply, many countries of the South are dismissing the concept of biological resources as the common heritage of humankind and are no longer willing to allow unconditional free access to the genetic resources under their national jurisdictions. While they concede that the international community has a common concern in conserving these resources, those countries maintain that they have the right to exploit and benefit from them.

Most industrialized countries on the other hand are refusing to make commitments on access to biotechnology and other associated technologies. But at the last meeting, some developing countries went as far as stating that they would not be willing to sign a convention which did not contain provisions related to the development and transfer of technology. Both issues of access – to genetic resources, and to technology – are being complicated by the extension of patents to cover life forms and the products of biotechnology, and by pressure on all countries to apply them. It is now agreed that the question of intellectual property rights should be reflected in the convention, taking into account the problems of developing countries, but it is unclear what form this will take. Clearly, a major North-South deal will be required on these issues to allow the negotiations to progress.

During the preliminary discussions, many delegations drew attention to the importance of local communities and their traditional technologies in conserving, developing and utilizing biodiversity. They agreed that there is a need to reward and sustain informal innovation by local people and their associated knowledge. The importance of biotechnology to meet the needs of small farmers was also emphasised. But so far, there has been little action on this issue. There is an urgent need to find mechanisms to put these good intentions into practice. Involvement of communities in strategies for conservation and sustainable development must be the very basis of the convention, not merely added on as an afterthought.

Vital components for a meaningful agreement

We now concentrate on the three inter-related issues which are both key to making progress in the negotiations and vital components of a meaningful agreement: access to genetic resources; technology transfer; and support for local communities.

1. Access to and control over genetic resources.

> *" There are various instances where transnational corporations have exploited the rich genetic diversity of developing countries as a free resource for research and development. The products of such research are then patented and sold back to the developing countries at excessively high prices. This must cease. We must formulate mechanisms for effective cooperation with reciprocal benefits between biotechnologically-*

rich developed countries and gene-rich developing countries...".

Malaysian delegate to the UN General Assembly, November 1990.

The current legal status of genetic resources is unclear at best. It is based upon a poorly defined concept of "common heritage" and the corollary principle of "free access" to genetic resources. But these principles are increasingly being undermined by restrictions imposed on the exchange of genetic resources. On the one hand there are political and economic embargoes being placed on germplasm exchange. But more far-reaching in effect are patents and other intellectual property rights now being granted on plants and animals and their genetic characteristics. The result is that genetic resources in the fields and forests of the South are normally freely available without charge, while those based on them and developed in the North become subject to monopoly control by private companies. As a reaction to this unbalanced situation, the developing countries are beginning to stake governmental claim to genetic resources in their territory, and to restrict access to them.

From this confused and biased situation, two different approaches are evolving in separate directions within two UN fora: the Commission on Plant Genetic Resources in FAO, and the negotiations towards the biodiversity convention under UNEP.

In FAO, the approach is based on the International Undertaking on Plant Genetic Resources, now signed by over 90 countries. Its objective is "..to ensure that plant genetic resources (...) will be explored, preserved, evaluated and made available (...) based on the universally accepted principle that plant genetic resources are a heritage of mankind and consequently should be available without restriction". Thus, in this forum, the principle of free exchange is endorsed by the majority of states, both developed and developing. They elaborate on this principle however, stating that free access does not necessarily mean free of charge, and that the rights of farmers should be recognized and rewarded, based on "the enormous contribution that farmers of all regions have made to the conservation and development of plant genetic resources." So far, however, Farmers' Rights remains an unelaborated concept; concrete measures to benefit farming communities are lacking, and the necessary funding has not been forthcoming.

During the preparations for the biodiversity convention, organized by UNEP, many representatives from the same governments party to the International Undertaking have been taking a different approach: rather than revising the present system of common heritage and free access, they are beginning to reject it altogether. At their last meeting in Nairobi, the working group of legal and technical experts on biological diversity concluded "that the heritage of mankind should not be included in the convention," and "it was largely agreed that accessibility to biological diversity should be based on mutual agreement between countries. Several countries objected to the concept of free access as such (...)".

Many have suggested that the International Undertaking be a protocol to the convention, perhaps after modification in order to include genetic resources other than from plants. Clearly, if these two initiatives are to be brought together, the fundamental inconsistency on access has to be resolved. The overall objective must be to balance access with remuneration to germplasm donors, and recognition of both the interests of local communities and of national sovereignty over natural resources. This should be negotiated in the framework of the proposed convention.

Ninety percent of our global food resources are based on just 30 species, so the continued exchange of genetic resources is vital to food security. Given the way in which crops have spread throughout the world and away from their centres of origin, no region can afford to isolate itself – or be isolated from – access to plant germplasm in other regions of diversity. This is more important for the industrialized nations, comparatively poor in biological wealth, but ultimately applies to all. In addition to providing the basis of our food supply, yet to be discovered properties of plants will provide the global community with valuable medicines and other products. In resolving the differences between member states in FAO and UNEP therefore, it is vital that access to genetic resources is guaranteed.

This need not conflict with the rights of sovereign states. Access should be unrestricted, but not unconditional or free of charge. The negotiated agreement in the framework of the convention could provide mechanisms to ensure that there is equitable remuneration to germplasm donors, that access can be regulated by states, and that intellectual property rights in the North do not restrict access to genetic resources contained within new plant varieties and the products of biotechnology. The benefits of guaranteed access would be limited to parties to the convention, who would also have to meet the obligations. For the industrialized countries this would include mandatory funding.

Thus, the agreement might include the following components:

- While the authority to grant access to genetic resources would rest with national governments, parties to the convention would refrain from imposing undue restrictions on access, and would be obliged to cooperate with other parties to allow their access.

- Access would be regulated to ensure that the resource base is not damaged by collection exercises, that the rights and needs of local communities are respected, and that collectors provide a share of the benefit.

- Developed countries would ensure that patents, or other IPR systems, do not create restrictions on free access to genetic resources.

- Contributions to a Fund for Biodiversity Conservation and Use would be mandatory, with the bulk coming from industrialized countries. This would ensure remuneration for the conservation and development of genetic resources, particularly by local communities.

The funding mechanism, and other measures in the convention, should also promote access to the new technologies, and ensure support for local communities.

2. Technology transfer to benefit the poor.

> *"Industrialised countries need to assist developing countries in their environment-related efforts, inter alia by the provision of additional financial resources ... also by facilitating access to, and transfer of, environmentally sound technologies, on a fair and most favourable basis."*
>
> European Community statement to UN General Assembly, November 1990

The developing countries have been arguing for access to the new technologies which are generated mainly in the industrialized countries. Providing remuneration to the South for genetic resources used as raw materials for the North's biotechnology industries is all well and good, but the South should benefit more directly from its immense resource base. In the context of biodiversity, the new biotechnologies are particularly relevant.

However, transfer of technologies from developed to developing countries can have negative -as well as positive- effects on socio-economic development and on conservation. The Green Revolution which introduced "High-Yielding Varieties" into the farming systems of the South is a case in point. The new technology might have been successful in increasing production of some major crops in some regions, but it also increased income disparities by accentuating existing inequalities. Thus it did little to alleviate poverty amongst the sectors of society most in need. It also increased dependence on external sources of fertilizers, pesticides, and knowledge and contributed enormously to genetic erosion by displacing the traditional varieties which form the biological resource base for food security.

When the livelihoods of local communities are undermined by technological change, their capacity to manage and protect their environment are also lost.

The socio-impact of the new biotechnologies is likely to be even more profound. These new techniques are being developed predominately for and by the industrialized countries where most of the research is under the control of the private sector. Developed in such conditions, these new technologies are already having negative implications for many Third World countries. Biotechnology companies are creating substitutes for agricultural products traditionally exported from the South, threatening to deprive developing countries from much of their desperately needed export earnings. The little work done on tropical crops is mostly focusing on high input plantation crops to the detriment of small scale farmers.

The South needs greater access to and control over these new technologies so that it can profit more directly from its own resources, diversify production, and use the technologies for the benefit of its people. If biotechnology is continued to be developed exclusively in the North, the technological gap will widen further still.

New mechanisms are required to promote the availability of technology in developing countries. But technology transfer should not be understood simply as putting the current biotechnology approach into functioning in the South. The mechanisms must have clear objectives: to reduce rather than increase dependence on developed countries; to build upon local knowledge and traditional technologies rather than displacing them; and to be based upon the real needs of the poor. Therefore, the convention should include the following principles:

- **Making technology available.** The convention should include mechanisms to promote access to biotechnological tools and other techniques vital for the conservation and use of biological resources. These might include modifications to intellectual property legislation, and other incentives to encourage technology transfer. Programmes will be required to ensure not only the transfer of relevant technologies but also to integrate them into local research systems. In particular, programmes will be needed to strengthen and better focus current agricultural research in many developing countries and International Centres.

- **Capacity building.** Support is required to build indigenous capacities in research and development. Such support would include institution building, access to relevant scientific and technical information, training, and the protection and development of already existing technologies. The International Agricultural Research Centres could have an increased capacity building function. Special attention should be given to improving the regulatory framework, so that developing countries can enforce appropriate safety measures.

- **A participatory approach.** Local communities should be involved in setting the priorities for research and researchers must learn from existing indigenous technologies, customs and constraints. To these ends, measures are required to ensure an effective flow of information. Since linking biotechnology research to the needs of poor farmers is especially difficult given the gulf between the molecular sophistication of the biotechnologist and the cultural sophistication of the farmer, farmers organizations, other NGOs, and local government agencies should be supported and involved in this process.

- **Funding for public research.** Most biotechnological research is conducted by private companies who focus on large markets which provide necessary profits but are unlikely to coincide with the particular needs of local communities. To redress the balance, funding is required for appropriate and well-planned public research into crops and products of biotechnology likely to have a high local socio-economic importance but do not provide global markets.

- **Ensure sustainability.** New techniques for the conservation of biological resources should be developed and used. More important still, are measures to ensure that biological resources are utilized sustainably by the new technologies. This requires *inter alia* a research effort to support and develop agricultural and other traditional production systems which are based on the maintenance and use of biological diversity.

As the importance of technology for development and environmental protection grows, it will become increasingly necessary that a deal on technology transfer is struck. Great care will be required to make sure that it benefits communities in the South and supports rather than undermines their conservation efforts.

3. Rights and incentives for local communities.

> *"If you want to rehabilitate the environment, you must rely on villagers and not on government officers to do the job. But people will care for their environment only if they have legal rights to manage it and to use its products. People already have the knowledge, what they must get are rights over their local environments."*
>
> Anil Agarwal, Centre for Science & Environment, New Dehli, 1990.

The contribution of farmers and other local communities in protecting and developing biological resources - in particular a diversity of cultivated, semi-wild and wild plants used for food, fuel and medicines - has been formally recognized by the international community. By now, "Farmers' Rights" are talked about at all relevant international conferences. But farmers are yet to benefit from these rights. The negotiations for a convention and its associated funding mechanism provide an opportunity to attach tangible benefits to the notion.

Local communities must be allowed to assume a major role in managing and benefitting from the resources on which their livelihoods and cultures depend. Currently, research and extension practices tend to bypass the needs of local communities, and their organizations are often under threat by several forces in many countries. Rather than strengthening a "top-down" approach, the convention should facilitate a decentralized, participatory approach to conservation, by providing dynamic support to local communities. This is crucial to maintain diversity and to promote equity and development.

The convention should contain a package of legal, financial, institutional and technical measures to support local communities. These should include the following:

- **Protecting and promoting local knowledge and innovations.** The value of local innovations, associated knowledge and production systems based on biological resources should be recognized and given legal standing to prevent abuse. Local knowledge should be promoted in educational programmes to improve public awareness of its value. The important role of women, who often hold precious knowledge about traditional food and medicinal plants and their uses, should be recognized and given due support. Where biological material and associated knowledge is collected by external agents, local people should be involved as equal partners and should be provided a share of the benefits from the resources and information collected.

- **Support conservation and utilization of local biological resources.** Local communities should receive direct and personal training in plant genetic resource conservation techniques applicable to local varieties, including breeding methods and seed supply systems. Grants and credits should be made available to promote the utilization and improvement of local varieties including, for example, their marketing and processing.

- **Strengthen local-based research and development.** Efforts should be made to ensure that communities have access to appropriate research and extension services. Research and extension should be reorientated towards a participatory, "farmer-first" approach, which addresses the needs of small scale and especially women farmers. Funding should be made available for grassroots initiatives, perhaps through local farmers' organizations. The links between farmers groups, other NGOs and national organizations should be strengthened. National and International research programmes should redirect their research efforts to concentrate on crops and farming systems relevant to small farmers.

- **Support community institutions and improve security.** Village institutions should be supported and encouraged, and the role of non-governmental organizations promoted, particularly with respect to providing training and infra-structural support for group work. Common property management institutions could be reestablished where they have proven effective in the past. Improved land tenure systems and other ways of increasing security of local peoples should be investigated and promoted where possible.

Successful policies to promote conservation and use of biological resources by local communities will have a double benefit. Firstly, their efforts in maintaining and developing food crops, medicinal plants and their wild and semi-wild relatives makes a direct and vital contribution to practical conservation of the world's biological resources. Additionally, such communities form a large part of a growing population; sustainable development of their systems of production is the key to improving food security, reducing poverty, and reducing its consequential effects of environmental degradation in large parts of the world.

Conclusion

If a comprehensive agreement is to be reached in the framework of the proposed convention, then the current emphasis in the negotiations on actions at the national level must be fully complemented by a strengthening of local and international actions.

Internationally, the question of access to genetic resources and to related technology has to be resolved. This may well prove to be where a major North-South deal takes place. The political momentum of the process towards the 1992 Conference on Environment and Development should be used to secure such a global agreement.

But both global agreements and national measures must be rooted in the needs of rural communities at the local level who are the real actors in development of biological resources. Effective conservation and development can not take place unless their needs are met.

Note: This article is from "Briefing on Biodiversity ", N0.2 , February 1991, prepared by GRAIN, Apartado 23398, E-08080 Barcelona, Spain.

APPENDIX

Outline of the Convention

Note: ● *indicates a summary of an element which has a wide degree of support*
➤ *indicates a summary of an element which is more controversial*

Chapter I: PREAMBLE/OBJECTIVES

● objective: to conserve biodiversity for the benefit of present and future generations
➤ objective: to establish an equitable system of measures for conservation and sustainable use and to acknowledge the special situation of developing countries
● identifies habitat destruction, pollution and 'unsound management' as the major causes of biodiversity loss
● recognizes national sovereignty and the role of local peoples

Chapter III: FUNDAMENTAL PRINCIPLES

➤ "common heritage" or "common interest"
➤ free access or access restricted by states
● addresses conservation and utilization
● equitable sharing of costs and benefits

Chapter III: GENERAL OBLIGATIONS

● duty to take appropriate conservation measures
● duty to refrain from actions harmful to biodiversity in other states or beyond national jurisdiction
● duty to promote public/community awareness
● mutual assistance

Chapter V: MEASURES FOR CONSERVATION AND SUSTAINABLE USE (AT NATIONAL LEVEL)*

***in situ* and *ex situ* conservation**

● surveys, information gathering, identification of important areas
● protection of important areas
● establish centres for *ex situ* conservation
● recovery and rehabilitation of threatened species, habitats and ecosystems
● reduction of pollution

sustainable use of wild resources

- support land use activities compatible with the maintenance of biodiversity
- recognition of the important role of biotechnology and of its potential dangers
- need to recognize and maintain local information on biodiversity and to reward and sustain informal innovation by local people
- integrate conservation into national development plans, ensure utilization does not have adverse impacts
- ➤ need for environmental impact assessments for proposed developments

research, training & education

- train taxonomists, ethnobotanists, specialists in sustainable agriculture and ecosystem function
- promote research in all these areas
- formal and non-formal education of public and of local populations
- capacity building in research institutions of developing countries
- strengthen information exchange between governmental and non-governmental agencies

Chapter VI: AVAILABILITY OF TECHNOLOGY AND ACCESS TO BIOLOGICAL DIVERSITY**

- guarantee availability and access to biological diversity for research, training, surveying & monitoring
- ➤ guarantee availability and access to biological diversity for other purposes
- ➤ preferential treatment for countries of origin
- ➤ guarantee availability and access to technology & information
- ➤ rules for patents and property rights on genetic resources

Chapter VII: TECHNOLOGY TRANSFER**

- ➤ obligation on industrial states to promote
- ➤ technology transfer technology transfer on a non-commercial or preferential basis
- ➤ undertaking for legislation for obligations on private companies
- ➤ promote exchange of information
- scientific cooperation and cooperation in training of experts
- promote joint ventures in biotechnology

- funding to facilitate technology transfer
- a clearing house mechanism to facilitate technology transfer

Chapter VIII: TECHNICAL ASSISTANCE**

- obligation to provide reasonable levels of technical assistance
- a clearing house mechanism to facilitate provision of technical assistance

Chapter IX: FINANCIAL MECHANISMS

- need for genuine additional resources
- mechanisms to be elaborated
  - ➤ create new fund or adapt existing funds
  - ➤ explore Montreal Protocol and World Bank's GEF as models

Chapter X: RELATIONSHIP WITH OTHER AGREEMENTS

- not to replace other agreements
- other agreements to be supported
- other agreements eligible to become protocols

Chapter XI: INSTITUTIONAL MEASURES AT THE NATIONAL LEVEL*

- obligation to legislate to set up specialised institutions at the national level

Chapter XII: INTERNATIONAL COOPERATION*

- promote international cooperation

Notes

1.Chapter II contains definitions; Chapter XIII - institutional arrangements for the Convention; Chapter XIV - dispute settlement mechanisms; Chapter XV will deal with annexes, protocols and amendments.

2.It has been proposed that the structure of the convention be reorganized, with chapters V and XI, marked *, forming a new chapter on measures at the national level, and chapters VI - VIII and XII, marked **, forming a new chapter on international mechanisms.